THE MARS PROJECT

T0321431

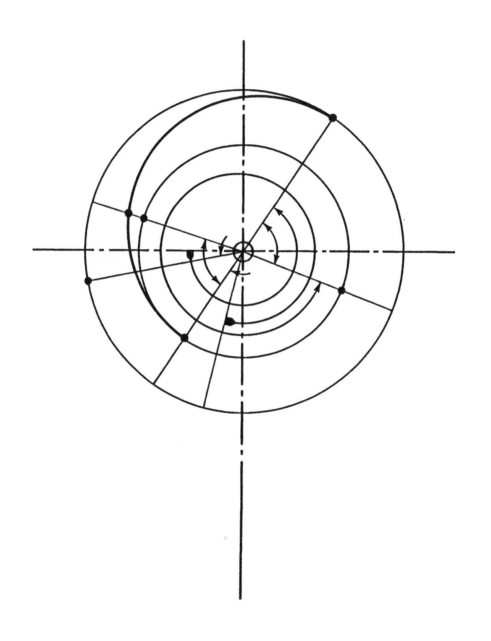

Wernher von Braun

THE MARS PROJECT

Foreword by Thomas O. Paine

UNIVERSITY OF ILLINOIS PRESS   URBANA AND CHICAGO

First Illinois paperback, 1991
© 1953, 1962, 1991 by the Board of Trustees of the
University of Illinois
German edition © 1952 by Bechtle Verlag, Esslingen, Germany
Manufactured in the United States of America
P 8 7 6 5

⊚ This book is printed on acid-free paper.

Library of Congress Cataloging-in-Publication Data
Von Braun, Wernher, 1912–1977
[Mars project. English]
The Mars project / Wernher von Braun ; foreword by
Thomas O. Paine.
p.   cm. — (Illini Books edition)
Translation of: Das Mars project.
Reprint. Originally published: Urbana : University of
Illinois Press, 1953.
ISBN 0-252-06227-2 (alk. paper) / ISBN 978-0-252-06227-8
1. Space flight to Mars. 2. Space ships. I. Title.
TL799.M3V62   1991
629.45'53—dc20   90-27646
                                                  CIP

CONTENTS

The algorithm of spaceflight laid out step-by-step in the terse lines of Wernher von Braun's *Mars Project* displays the logic that seventeen years later carried astronauts to the Moon. Humans have always dreamed of travel to other worlds. The great rocket pioneers—Tsiolkovsky, Oberth, Goddard, Tsander, von Braun, Korolev, and others—were inspired by the prospect of interplanetary voyages. They sold (and oversold) other applications of rockets, but their real motivation was always spaceflight.

From boyhood, Wernher von Braun envisioned voyages to other worlds. He once told me that it was the gift of a telescope that turned his young eyes skyward and pointed his career toward the stars. As a teenager in pre–World War II Berlin, he joined a group of enthusiastic amateurs designing innovative systems and defining technical breakthroughs required for interplanetary flight. While pursuing his engineering education, he applied his expanding knowledge to the development of critical components for liquid fuel rockets. Recurring, spectacular explosions punctuated these pioneering experiments.

As war clouds gathered, the innovative young engineer was recruited by Captain Walter Dornberger, a thirty-five-year-old artillery officer ordered to build long-range military rockets in lieu of the aircraft prohibited to Germany by the Treaty of Versailles. Despite his youth, von Braun soon became the technical leader of the group and proposed moving the growing enterprise to Peenemünde, an island in the Baltic Sea where his father had hunted ducks.

The space age can be said to have begun on October 3, 1942, with the flight of von Braun's first A-4 (V-2) missile. This 46.1-foot–high, single-stage rocket with a 2,200-pound payload was propelled at 3,500 miles an hour for 200 miles by an alcohol–liquid oxygen engine capable of developing 56,000 pounds of thrust. V-2 bombardment of London was throttled by Allied armies invading Germany, but not until 1,054 rockets had struck England between September 8, 1944, and March 27, 1945. Meanwhile, von Braun survived fleets of Allied bombers that devastated the test complex; he also survived arrest by the Gestapo for defeatist statements about Germany's chances of winning the war. Charged with advocating the building of interplanetary spacecraft instead of military weapons, he spent two weeks in a prison cell in Stetten in March 1944.

In February 1945, von Braun fled Peenemünde ahead of the advancing Red Army. He led his battered rocket team southwest with crates of rocket data; on May 2, 1945, they surrendered to advancing American troops near Reutte, Austria. Finding the German team remarkably cooperative, the U.S. Army transported 115 of the captured experts and 100 V-2s to New Mexico to continue rocket development and high-altitude research. Von Braun, like Moses, led his expatriates through the desert toward a distant promised land.

In the course of his subsequent experimental work, von Braun took a fresh look at interplanetary flight based upon his rocket team's cumulative experience in Germany and the United States. Ten years after the first V-2 rocket flight, he published his classic *Das Marsprojekt* in a special issue of the magazine *Weltraumfahrt*. This work also appeared in 1952 as a slim volume, *Das Marsprojekt: Studie einer interplanetarischen Expedition*, which was translated and published in 1953 as *The Mars Project*, which in turn stimulated a series of popular articles in *Collier's* magazine. Chesley Bonestell's dramatic illustrations of future space shuttles, space stations, astronaut-tended space telescopes, and interplanetary spacecraft voyaging to Mars inspired a generation of young people to technical careers that could help make spaceflight a reality.

Von Braun's seventy-person Mars expedition included a fleet of forty-six space shuttles of 39-ton lift capacity (NASA's space shuttles lift 20 tons to orbit). With a turnaround time of 10 days (NASA's shuttles require 75-125 days), these reusable vehicles could make 950 flights to orbit in eight months, allowing for six vehicles being continually out of service. This would require 5.32 million tons of fuel costing around $500 million, which von Braun equated to ten times the high-octane aviation gasoline burned in the six months of the Berlin airlift. The result would be ten fully fueled spaceships, each weighing 3,720 metric tons, ready to depart Earth's orbit in the plane of the ecliptic on a 260-day voyage to Mars.

While von Braun's team was working in the United States, Helmut Gröttrup and his engineers were transferring V-2 technology to Soviet teams led by Sergei Korolev, Valentin Glushko, and others. Stalin was particularly impressed by Eugen Sänger and Irene Bredt's plans for an antipodal bomber capable of attacking America; this conceptual design of an aerospace plane resembled a huge piloted V-2 with wings. He directed that the highest priority be given to intercontinental ballistic missile (ICBM) development and atomic bombs. New launch complexes were built at Kapustin Yar and Tyuratam to test increasingly powerful Soviet rockets. Impelled by technical advances and the intensifying cold war, ICBM development went into high gear in 1954 on both sides of the Iron Curtain. The resulting advances in rocketry led scientists organizing the International Geophysical Year (IGY) to propose that artificial satellites be launched in 1957.

Moscow's response to the IGY proposal was the Commission for Interplanetary Transport (ICIC) within the Soviet Academy of Sciences. Led by the academician Leonid Sedov, ICIC's bold mission was to develop robotic spacecraft for interplanetary flight. On July 29 and 30, 1955, both Washington and Moscow announced plans to launch satellites during the IGY. The navy's Project Vanguard carried the banner for the United States; Sergei Korolev led the Soviet ICBM/Satellite Launcher Program. The latter's team successfully flew the first R-7, *Semyorka* ("Good Old Number Seven"), two years later, on August 3, 1957.

On October 4, a Soviet R-7 launched the 184-pound *Sputnik* into orbit. Americans were shocked by this spectacular achievement, but they should not have been; not only had Moscow announced its intentions, but von Braun's U.S. team had been ready to launch a small satellite since 1956 (its proposal was shelved in favor of the navy project). On November 3, Moscow celebrated the fortieth anniversary of the Russian Revolution in spectacular style by rocketing the 6-ton *Sputnik II* into orbit. The payload included an 1,121-pound capsule with geophysical equipment, telemetry, and a life-support system for the canine cosmonaut Laika, whose presence clearly presaged human spaceflight. In response to American charges that German experts were behind the Soviet achievements, Nikita Khrushchev smilingly pointed out that the United States had most of the experts and then asked why von Braun's team was not able to launch an American satellite.

A nationwide television audience watched the U.S. Navy's heralded *Vanguard* rocket explode and collapse on the launch pad on December 6. This embarrassing fiasco, after dazzling Soviet achievements, prompted Washington to give the eager von Braun and his team the green light to launch a satellite with their *Jupiter C* rocket. On January 31, 1958, America's 10.5-pound *Explorer I* soared into orbit with two micrometeoroid detectors, a Geiger counter, and telemetry. At less than 1 percent of the weight of *Sputnik II*, the miniaturized instruments on board nevertheless returned more valuable scientific information by discovering and mapping the Van Allen radiation belt that surrounds Earth.

On the advice of President Eisenhower's Science Advisory Committee, and after a thoughtful review of alternative courses, on April 14 the U.S. Congress passed the National Aeronautics and Space Act of 1958 (S. 3609; H.R. 11881); this farsighted piece of legislation created the civilian NASA. America's fledgling space agency organized itself around the predecessor National Advisory Committee for Aeronautics, the Naval Research Laboratory's Vanguard team, and two groups transferred from the army: von Braun's Redstone Arsenal team at Huntsville, Alabama, and the Jet Propulsion Laboratory at Pasadena, California. In the Soviet Union, rapid progress continued as new payloads weighing up to 6,500 pounds were launched. Khrushchev sneered that America would have to launch a lot of orange-sized sputniks to catch up.

On April 12, 1961, Yury Alekseyevich Gagarin blazed a human trail into orbit aboard Sergei Korolev's *Vostok I* ("The East"). His dramatic spaceflight captured the imagination of the world and called into question American technology and leadership. The Kennedy administration, smarting under Fidel Castro's success at the Bay of Pigs, resolved to gain the lead in space and explored three alternative programs to achieve this goal. An orbiting space station was rejected as too easily within Soviet capabilities, and an expedition to Mars was judged too difficult to accomplish within a decade. A landing on the Moon appeared to be an achievable project that would challenge NASA in all areas of spaceflight and establish America as the preeminent spacefaring nation.

The projected $20 billion cost of a lunar landing ($70 billion in 1990 dollars) would boost NASA's peak 1965 budget to 0.78 percent of the gross national product (GNP), but the alternative of surrendering space leadership appeared unthinkable. Four months after Gagarin's flight, the Berlin Wall was erected, while Red Army tanks patrolled Eastern European capitals and the Soviet Union's shoe-pounding premier threatened at the United Nations to bury the West. Washington saw a threat to world peace from military adventurism by Kremlin leaders miscalculating the relative technological strengths of the superpowers. Although no American had yet flown in orbit, on May 25, 1961, President Kennedy asked a cheering Congress to direct NASA to land astronauts on the Moon within the decade.

The national goal of a lunar landing within eight years challenged the U.S. aerospace enterprise across the entire spectrum of technologies. NASA administrator James E. Webb drew from government, industry, and university circles to create a superb management team that operated on a semiwartime footing. Ninety percent of the Apollo budget was spent outside the space agency as 400,000 Americans across the country were attracted to NASA's open program and inspiring goals.

Von Braun led Eberhardt Rees, Kurt Debus, and other key Peenemünde engineers in a fast-paced project to develop the essential heavy-lift launch vehicle: a giant three-stage, 363-foot rocket called the *Saturn V*. The first stage of this unprecedented booster developed 7.5 million pounds of thrust from five mighty F-1 kerosene–liquid oxygen engines burning 15 tons of fuel per second (the fuel pumps alone had greater horsepower than the turbines driving the new ocean liner *Queen Elizabeth*). The two liquid hydrogen–liquid oxygen upper stages lifted 120 tons of payload into orbit for the 240,000-mile voyage to the Moon. NASA's conceptual systems design group adopted an innovative Lunar Orbit Rendezvous concept that substituted electronic docking prowess for brute rocket power. A giant new launch complex was built at Cape Canaveral, a new manned spaceflight center was constructed at Houston, a worldwide tracking network was created, and new industrial and university research facilities were established across the country.

As the end of the 1960s approached, precursor robotic missions were launched to characterize the lunar surface, and every spaceflight system and component was tested and retested. The impetus of Project Apollo's purposeful activities spurred many parallel developments, from Mariner spacecraft missions to Venus and Mars to the creation and spin-off of valuable global weather and communications satellite systems.

On July 20, 1969, Neil Armstrong, Buzz Aldrin, and Mike Collins flew the historic *Apollo 11* mission that touched down on the lunar Sea of Tranquility—on time and within budget. Their footprints on the Moon's ancient surface record humanity's first steps toward a multiplanet civilization. National jubilation and worldwide acclaim greeted America's triumph. By initiating human exploration of the Moon through NASA's open civilian space program, Presidents Kennedy and Johnson and congressional leaders had made the United States the preeminent spacefaring nation. To von Braun, this achievement marked the next step in the evolution of life. He equated astronauts crossing space to explore the Moon to the first marine life learning to live on land. *Apollo 11* was a boyhood dream come true, a beacon lighting the way to our future as a multiplanet species.

Because of his outstanding ability to envision the future, I asked von Braun to join me at NASA headquarters in Washington to help plan America's post-Apollo program. In 1969 President Nixon appointed a Space Task Group to explore manned spaceflight alternatives, including a large orbiting space station, continuing lunar exploration, and a long-range mission to Mars. Von Braun contributed to all these plans but none were pursued; the "Moon Race" was won, and national attention had turned elsewhere. The divisive Vietnam conflict made high-tech programs suspect, and science education came to be seen as elitist. With no future U.S. manned mission in prospect, *Saturn V* production was terminated and the space program slumped back to a third of its 1960s peak. At the same time American universities experienced a steady decline in young people pursuing graduate work in science and technology.

Dissatisfied with NASA's aimlessness, in 1985 Congress created the Presidential National Commission on Space to look thirty years into the future and recommend long-range goals for America's civilian space program. The commission's final report, *Pioneering the Space Frontier*, proposed to the president and Congress a balanced, future-oriented program. The overarching recommendation was that America "lead the exploration and development of the space frontier, advancing science, technology and enterprise, and building institutions and systems that make accessible vast new resources and support human settlements beyond Earth's orbit, from the highlands of the Moon to the plains of Mars."

This was the goal anticipated by von Braun's classic *Mars Project.* On the twentieth anniversary of the first lunar landing,

President Bush delivered a historic address at the Smithsonian Air and Space Museum. Standing before the Wright brothers' 1903 *Flyer*, Lindbergh's 1927 *Spirit of Saint Louis*, and the 1969 spaceship *Columbia* in which *Apollo 11* astronauts flew to the Moon, the president directed NASA to prepare plans for an orbiting space station, lunar research bases, and human exploration of Mars. The 500th anniversary of Columbus's discovery of the new world will see that world setting sail for other new worlds across the ocean of space.

Von Braun watched the first humans explore the Moon; and he knew that among our children are the first explorers of Mars. As interplanetary travel becomes increasingly feasible and affordable in the twenty-first century, the expansion of life outward from its earthly cradle will become an enduring international goal. Space exploration and settlement will be accelerated by exponentially growing world economies, decreasing superpower confrontation, continuing advances in science and technology, and advancing spaceflight experience. Human intelligence is destined to activate the evolution of life on other worlds.

It is thus fitting that I close with Wernher von Braun's clear vision of the next century:

Only a miraculous insight could have enabled the scientists of the eighteenth century to foresee the birth of electrical engineering in the nineteenth. It would have required a revelation of equal inspiration for a scientist of the nineteenth century to foresee the nuclear power plants of the twentieth. No doubt, the twenty-first century will hold equal surprises, and more of them. But not everything will be a surprise. It seems certain that the twenty-first century will be the century of scientific and commercial activities in outer space, of manned interplanetary flight, and the establishment of permanent human footholds outside the planet Earth.

Thomas O. Paine

Most of the calculations which form the substance of this book were made in 1948. The author did the work in his spare time and his sole computational tool was a slide rule. During the intervening fourteen years flight through outer space, then still a most controversial topic and at best a subject of cautious feasibility studies, has captured the imagination of the entire human race and has developed, in this country alone, into a multibillion-dollar-a-year research and development program.

It would be disappointing, indeed, had these fourteen years of determined efforts not drastically raised the general level of techniques and know-how on which up-to-date plans for an expedition to Mars could be based. My basic objective during the preparation of *The Mars Project* had been to demonstrate that on the basis of the technologies and the know-how then available (in 1948), the launching of a large expedition to Mars was a definite technical possibility. I hope the calculations proved my point, but they also indicated that the undertaking would be quite costly. "The logistic requirements for a large elaborate expedition to Mars," I said in the introduction, "are no greater than those for a minor military operation extending over a limited theater of war." I am now ready to retract from this statement by saying that on the basis of technological advancements available or in sight in the year 1962, a large expedition to Mars will be possible in fifteen or twenty years at a cost which will be only a minute fraction of our yearly national defense budget.

What are the most important technological breakthroughs that have brought the planet Mars so much closer in our reach?

The greatest single advance is probably the availability of reliable rocket engines burning liquid hydrogen and liquid oxygen. These engines reach an exhaust velocity (a yardstick for rocket motor thrust obtained for a given rate of propellant consumption) of 4,200 m/sec versus the mere 2,800 m/sec of the hydrazine-and-nitric-acid engines which power both the orbital ferry vessels and the deep-space ships in *The Mars Project*. The tremendous performance gains obtained by the use of liquid hydrogen—along with structural and other refinements—are best illustrated by a comparison. Take the huge ADVANCED SATURN C-5 rocket ship, whose development our Huntsville Space Flight Center is presently conducting for the National Aeronautics and Space Administration's Manned Lunar Landing project. This two-stage-to-orbit rocket burns liquid hydrogen in its upper stage and will be capable of hauling into a departure orbit about five times as much payload (125 metric tons) as the hydrazine-powered three-stage ferry vessel described in this book. Moreover, whereas the latter is presented as an enormous vehicle of a take-off weight of over 12,000 tons, the take-off weight of an ADVANCED SATURN C-5 will be only about 3,000 tons.

Another breakthrough has occurred in the field of nuclear rocket power. When *The Mars Project* went to print I expressed doubt whether "within the next twenty-five years an atomic rocket drive will be able to compete in cost with chemical propulsion." To be sure, over half of that time has elapsed and we still do not have any flyable nuclear rocket engines. But I think we are now well on our way toward getting at least two promising types.

One is the thermal nuclear engine, wherein liquid hydrogen is pumped into a multitude of tubular channels leading through a high temperature reactor. The emerging hot hydrogen gas is then expanded through a supersonic exhaust nozzle like a normal rocket jet. This engine promises to yield an exhaust velocity in the order of 9,000 m/sec—more than twice the performance of chemical hydrogen-oxygen engines and well over three times as much as the hydrazine-and-nitric-acid engines on which *The Mars Project* has been based.

Another type is the nuclear-powered ion engine. Here, the heat energy derived from the reactor is first converted into electricity. The electric power is then used to expel ionized (or electrically charged) atomic particles with the help of powerful electric fields. The expulsion (or exhaust) velocity of a good ion rocket engine is on the order of 100,000m/sec. Unfortunately, this spectacular performance figure loses some of its attraction because of the severe weight penalty incurred by the equipment required for the conversion of heat into electricity under space conditions. Nevertheless, nuclear-powered ion ships designed for departure from an Earth orbit look better than any other scheme for the deep-space portion of a Mars expedition and they seem to offer truly drastic gains in interplanetary payload capability.

Finally, we have learned a great deal about the problems associated with the return of a vehicle from outer space, its re-entry into the atmosphere and ensuing deceleration through air drag. In *The Mars Project* I had rather conservatively assumed that the most one could expect in this regard was a rather heavy glider capable of re-entry from a relatively low orbit around the Earth. For the returning Mars ships this meant, of course, that they had to carry enough propellants to reduce their approach speed with the help of retro-rocket power from the hyperbolic velocity (near the vertex of the capture hyperbola) down to orbital speed. This is a particularly costly demand since prior to their consumption the propellants for this terminal maneuver must be carried all the way to Mars and back. They are therefore tantamount to a tremendous additional payload for the entire voyage.

Experience with recovered missile nose cones and satellite capsules has taught us a few tricks and it now appears certain that heat-protected capsules can be built which can not only return from low orbits, but which are capable of safe re-entry into the atmosphere at hyperbolic speeds. Thus the entire retardation energy from tangential hyperbolic re-entry into the atmosphere to zero-speed touchdown on the Earth's surface can be dissipated by aerodynamic drag and no terminal power maneuver will be needed at all.

Combine these three factors—more economical surface-to-orbit ferry rockets, nuclear-powered ion ships for the deep-space voyage, and no need for propellant-consuming retardation maneuvers at the end of the voyage—and it becomes obvious why the technical and logistic requirements for an expedition to Mars have enormously shrunk in size.

There are many other questions which back in 1948 could not be convincingly answered for lack of direct evidence, but which now, in 1962, have lost much of their sting. The lifetime of the electronic gear aboard our artificial satellites proves that the meteor hazard in outer space is not too serious a consideration. Explorer VII was still swamping us with telemetered data a year after its injection into orbit. Radio communication over interplanetary distances has verified the calculations and estimates presented in Chapter F of this book. (Messages from Pioneer V were received over a distance of up to 22 million miles.) Temperature control in all the many artificial satellites and deep-space probes has worked pretty much as expected.

Finally, man himself has entered space. The historic flights of our American astronauts and their Soviet counterparts have demonstrated beyond a shadow of a doubt that an artificial cabin environment can be provided in space vehicles in which men can live and efficiently perform. They have further proven that man can communicate, drink, eat, and sleep in space and that he can take over manual control in case of equipment failures. True, the longest space flight performed so far lasted only one day. Thus the entire complex of human factors in interplanetary voyages involving travel times of many months is still a wide open book. We still do not know the amount of shielding needed to protect interplanetary crews from the combined radiation hazard posed by cosmic rays, trapped radiation (Van Allen Belt), and solar flares. Maybe we will have to provide our Mars ships with shielded "storm cellars" where the crews have to seek shelter during some of those still-unpredictable solar proton outbursts during which the radiation level in outer space seems to increase manifold.

But for all the still-remaining questions we can herald the happy tidings that none of our rocket probes into outer space has produced even the tiniest piece of evidence of the existence of any fundamental barrier to human interplanetary travel. Fourteen years ago we thought it was, but now we know it: the road to the planets is open.

THE MARS PROJECT

INTRODUCTION

Soon after the publication of the first serious papers on space travel, a spate of fanciful tales appeared — purporting to show vividly just how an interplanetary voyage would be carried out. The central figure in these stories was usually the heroic inventor. Surrounded by a little band of faithful followers, he secretly built a mysteriously streamlined space vessel in a remote back yard. Then, at the hour of midnight, he and his crew soared into the solar system to brave untold perils — successfully, of course.

Since the actual development of the long-range liquid rocket, it has been apparent that true space travel cannot be attained by any back-yard inventor, no matter how ingenious he might be. It can only be achieved by the coordinated might of scientists, technicians, and organizers belonging to very nearly every branch of modern science and industry. Astronomers, physicians, mathematicians, engineers, physicists, chemists, and test pilots are essential; but no less so are economists, businessmen, diplomats, and a host of others. We space rocketeers of all nations (where permitted) have made it our business to rally this kind of talent around the standard of space travel, which, in the nature of things, is synonymous with the future of rocketry.

I believe it is time to explode once and for all the theory of the solitary space rocket and its little band of bold interplanetary adventurers. No such lonesome, extra-orbital thermos bottle will ever escape earth's gravity and drift toward Mars. The research and development which preceded, and was essential to, the attainment of a modest 250 miles of altitude is a sufficient guarantee of that statement. During this research and development it was learned very thoroughly how some ridiculously insignificant incident can jeopardize the results of years of preparation. No one with even the most primitive knowledge of the subject can possibly believe that any dozen or so men could build and operate a functional space ship, or, for that matter, survive interstellar isolation for the required period and return to their home planet.

But let us assume for a moment that such a small group of men were actually to land on Mars or some other planet. How could they probe its inner mysteries? To do so they would require means of transportation and some form of movable, pressurized housing if they were to explore any considerable portion of the planetary surface. While a man encased in an impressively clumsy pressure suit, walking importantly around the base of a space vessel, makes a fine and interesting figure in a lunar or Martian moving picture, it is unlikely that he will gather much useful data about the heavenly body on which he stands.

In 1492 Columbus knew less about the far Atlantic than we do about the heavens, yet he chose not to sail with a flotilla of less than three ships, and history tends to prove that he might never have returned to Spanish shores with his report of discoveries had he entrusted his fate to a single bottom. So it is with interplanetary exploration: it must be done on the grand scale. Great numbers of professionals from many walks of life, trained to co-operate unfailingly, must be recruited. Such training will require years before each can fit his special ability into the pattern of the whole. Aside from the design and construction of the actual space vessels, tons of rations, water, oxygen, instruments, surface vehicles and all sorts of expeditionary equipment will be required. The whole expeditionary personnel, together with the inanimate objects required for the fulfillment of their purpose, must be distributed throughout a flotilla of space vessels traveling in close formation, so that help may be available in case of trouble or malfunction of a single ship. The flotilla will coast for months on end along elliptical paths and will require intership visiting, necessitating the use of "space boats." Obviously an ample supply of spare parts and repair equipment cannot be omitted.

Many space-travel enthusiasts are under the impression that the use of chemical propellants would require such enormous masses of fuel that an interplanetary expedition would be impossible, due to the resulting restriction of payload. That is why the fictional histories of such trips so frequently refer to "atomic propellants" or "atomic rocket drives." Anything similar would, of course, considerably simplify the problem, and I do not propose to deny the possibility that nuclear energy may someday propel space vessels. We must beware of using the word "impossible" when we speak of technical developments. However, atomic energy all by itself does not constitute a reactive propulsion system and all theories on the application of atomic energy to space ships are as yet purely speculative. We do not advance our cause if we rely upon vague hopes of that order. I am still to be convinced that within the next twenty-five years an atomic rocket drive will be able to compete in cost with chemical propulsion.

The study which follows proposes to prove that we can thrust an expedition to Mars with conventional, chemical propellants. It will be on the grand scale and it will be expensive, but neither the scale nor the expense would seem out of proportion to the capabilities of the expedition or to the results anticipated.

The study will deal with a flotilla of ten space vessels manned by not less than 70 men. Each ship of the flotilla will be assembled in a two-hour orbital path around the earth, to which three-stage ferry rockets will deliver all the necessary components such as propellants, structures, and personnel. Once the vessels are assembled, fueled, and "in all respects ready for space," they will leave this "orbit of departure" and begin a voyage which will take them out of the earth's field of gravity and set them into an elliptical orbit around the sun.

At the maximum solar distance of this ellipse which is tangent to the Martian orbit, the ten vessels will be attracted by the gravitational field of Mars, and their rocket motors will decelerate them and swing them into a lunar orbit around Mars. In this they will remain without any thrust application until the return voyage to earth is begun.

Three of the vessels will be equipped with "landing boats" for descent to Mars's surface. Of these three boats, two will return to the circum-Martian orbit after shedding the wings which enabled them to use the Martian atmosphere for a glider landing. The landing party will be transshipped to the seven interplanetary vessels, together with the crews of the three which bore the landing boats and whatever Martian materials have been gathered. The two boats and the three ships which bore them will be abandoned in the circum-Martian orbit, and the entire personnel will return to the earth-orbit in the seven remaining interplanetary ships. From this orbit, the men will return to the earth's surface by the upper stages of the same three-stage ferry vessels which served to build and equip the space ships.

There are three subdivisions of the study, according to the principal phases involved:

a. Three-stage ferry vessels for communication between earth's surface and the orbit of departure.

b. Space ships to reach the circum-Martian orbit and return from it to the circum-tellurian orbit.

c. "Boats" to land on Mars and then return to the circum-Martian orbit.

An outline of requirements for such an expedition, in terms of equipment and propellants, is included. The last two subdivisions include the bases upon which the computations of propellant performance are founded. There is also a study of interplanetary radio. No search for optima is included in regard to any of the three phases of the voyage. For example, the ferry vessel computations are based on the use of three-stage rockets, which is the minimum number capable of attaining the orbit. This also simplifies the problems of construction and salvaging the drop-off stages. A multiplicity of stages would, however, permit considerable weight-saving.

The use of liquefied gases as propellants is purposely avoided. In theory, liquid hydrogen and oxygen would save a great deal of weight, but their low boiling points would immensely complicate the supply and logistic problem, probably to such a degree that no gain would result. Another drawback of liquefied gases is that they preclude the use of collapsible propellant tanks made of synthetics. Such collapsible tanks would greatly facilitate the construction of the interplanetary vessels in the circum-earth orbit, both with respect to the weight involved and to ease of stowage in the ferry vessels. Thus all computations of power plant performance for ferry vessels, interplanetary ships, and landing craft are based on hydrazine (N_2H_4) and nitric acid (HNO_3). These propellants remain in the liquid phase at normal temperatures. There are several concessions to practicality of this order. Nevertheless, the logistic requirements for a large, elaborate expedition to Mars are no greater than those for a minor military operation extending over a limited theater of war.

The study applies exclusively to the *mechanical* problems — in particular to the problems of celestial mechanics — relating to a voyage to Mars. The technical data applying to the interplanetary and orbital vehicles required for such a trip are therefrom derived. The study aims at substantiating the technical possibility of a voyage to Mars, but by no means pretends to represent the optimum plan. It will be obvious that the very nature of such a limited work on such a broad subject precludes its being anything but a relatively rough outline in which certain simplifying assumptions must be made so that a very complex proposition might be brought within focus.

One example of such simplification is the assumption of truly circular orbits for both earth and Mars, despite the fact that the orbit of Mars in particular has a marked eccentricity. Again, it is obvious that the space vessels will be subjected simultaneously to the gravitational influence of various asters, but it is assumed that the ships are affected only by that of the heavenly body predominatingly operative. Without this simplifying assumption, it would be necessary to solve laboriously a multiple body problem by numerical integration. This would frustrate any attempt to present the problem of celestial mechanics briefly and concisely. It should be noted that simplifications of this order, as applied, exert but little influence upon the major relationships and thus do not negate the positive conclusions, based upon the computed results, to the effect that the undertaking is technically feasible.

The study furthermore takes no cognizance of problems and questions affecting the *development* of space vessels and ancillaries. It is limited to required and attainable performances of such space craft when emerging from the development stage.

No attempt has been made to treat the difficult problem of interplanetary astronavigation. The latter may be approached along three major paths, as follows: ferry rockets en route to and from the orbit of departure can be tracked from the ground by radio and radar, even by optical means, and informed by radio of their positions. When not too remote from earth and Mars, interplanetary space vessels may get fixes by observing the times of occultation of selected fixed stars behind the rims of the planets referred to. Further out in interplanetary space it will be necessary to have recourse to the angular relationships between the planets and fixed stars to obtain a fix. When fixes thus obtained do not coincide with the pre-calculated track, corrective thrust periods will be required, and a liberal propellant allowance for this purpose has been provided in the performance computations for each type of space rocket.

Another and perhaps vitally important factor left undiscussed is the extremely involved problem-complex revolving around human capacity for withstanding the hardships incident to an interplanetary voyage. It is known that temperature-control problems in space rockets are far from insoluble and that we can provide space farers with breathable atmosphere at adequate pressure. We are certain that they will have oxygen, food, and water in abundance and that they will not freeze to death, nor burn, nor suffocate. What we do not know is whether any man is capable of remaining bodily distant from this earth for nearly three years and returning in spiritual and bodily health. Is it possible, for example, to devise some means of training prospective crews to withstand the spiritual hardships incident to interplanetary exploration? Will extended exposure to primary cosmic radiation produce radiation sickness or loss of sight? We are only familiar with cosmic radiation on and near our earth, and there is some question as to whether it becomes stronger at a greater distance from the sun. This may well be — although we do not know — because the sun may have a strong magnetic field which deflects protons and heavy nuclei of primary cosmic radiation more strongly the closer they approach the sun.

There is, too, the problem of the effect of weightlessness which exists along all unpowered extra-atmospheric trajectories. A space farer within a space rocket is subjected to the same celestial laws of motion, whether the movement occurs within the fields of gravity of one or more heavenly bodies. There can exist no differential which might attract him towards any wall of the capsule within which he lives. Though it is known that short periods of weightlessness have inconsiderable effects, if any, extended exposure to it may be different. Space stations such as the much discussed "artificial satellite" may be so designed as to rotate around their own axes and thus provide "synthetic gravity" for the comfort of their occupants. Will it be necessary to apply the same expedient to ships designed for exploring Mars? Must we consider linking two vessels together and rotating them about their common center of gravity as they drift thrustless through space? Or shall we provide the Mars flotilla with annular or dumbbell-shaped "gravity cells" capable of self-rotation? These would be unloaded into space after thrust cut-off: they would drift alongside the flotilla and be available throughout most of the 260 days of progress along the ellipse leading to Mars, and offer the space crews a few hours per day during which each man could "feel his weight."

As to meteors, we know that space rockets within a few thousand miles of earth will scarcely run a risk of colliding with anything but tiniest particles of meteoric grains. We also know that we can protect pressurized and inhabited nacelles as well as propellant tanks against such meteoric particles by armoring them with thin sheet metal of extremely light weight. Years of systematic observation of shooting stars provide reliable data for this assumption. There is also the possibility of building self-sealing propellant tanks, fashioned after those bulletproof gasoline tanks used in military aircraft.

But how much do we know about meteoric concentration occurring between the orbits of earth and Mars? It is quite definite that in the asteroid belt — beyond Mars's orbit — there is a huge quantity of different-sized meteors circling the sun. Whether the frequency of occurrence of small meteoric particles begins to increase on this side of the Martian orbit, we do not know.

The reader will discover a great number of open questions between the lines of this modest document. We believe that we have some of the answers, but there are many which still require solution. There is scarcely any branch of science which has no bearing upon interplanetary flight, and this little booklet will have achieved its objective if it stimulates some of its readers to find, in their own particular specialties, contributions which may fill out one or more of the many gaps still existing in the scientific extrapolation of a voyage to a neighboring planet.

Acknowledgments are due Mr. Krafft Ehricke, Dr. Hans Friedrich, Dr. Josef Jenissen, Dr. Joachim Mühlner, Dr. Adolf Thiel and Dr. Carl Wagner for their valuable contribution to the preparation of this book. I furthermore thank Messrs. Breidenstein, Bechtle, Loeser, Koelle, and Gartmann, who made the publication possible, and Mr. Henry J. White, who prepared the English translation.

The staff members of the University of Illinois Press to whom I am particularly grateful are Dr. Donald Jackson, editor, and Miss Mary Ruth Kelley, production manager.

A

THREE-STAGE FERRY VESSELS

1. COMPILATION OF MAIN DATA WITH SYMBOL MEANINGS

TABLE 1. ORBIT OF DEPARTURE

Period of revolution	T_1	2 hr
Orbital radius	$R_{S,E}$	8,110 km
Altitude above earth's surface	$y_{S,E}$	1,730 km
Orbital velocity	$v_{ci,1}$	7.07 km/sec

TABLE 2. FIRST STAGE

Thrust (at expansion to an exit pressure equal to the ambient pressure)	F_I	12,800 t
Take-off weight	$W_{0,I}$	6,400 t
Empty weight	$W_{E,I}$	700 t
Final weight (1st stage empty; 2d and 3d stages full)	$W_{1,I}$	1,600 t
Mass ratio	E_I	4
Ratio of empty weight of 1st stage to propellant weight	k_I	0.146
Propellant weight	$W_{P,I}$	4,800 t
Rate of propellant flow	$\dot{W}_I$	55.81 t/sec
Exhaust velocity	u_I	2,250 m/sec
Nozzle exit pressure	$p_{e,I}$	0.7 kg/cm^2
Nozzle exit area	$A_{e,I}$	224 m^2
Initial acceleration (absolute)	$a_{0,I}$	18.64 m/sec^2
Take-off acceleration (relative)	$a_{to,I}$	8.83 m/sec^2
Final acceleration (absolute)	$a_{1,I}$	87.0 m/sec^2
Burning time	t_I	84 sec
Cut-off altitude	y_I	40 km
Cut-off velocity	v_I	2,350 m/sec
Cut-off distance	x_I	50 km
Angle of elevation at cut-off	φ_I	20.5°
Distance between take-off and landing (1st stage)	$x_{L,I}$	304 km
Length of 1st stage (without fins)	L_I	29 m
Diameter of 1st stage	D_I	20 m

Note: 1 t = 1 metric ton = 1,102 short tons

TABLE 3. SECOND STAGE

Thrust	F_{II}	1,600 t
Initial weight	$W_{0,II}$	900 t
Empty weight	$W_{E,II}$	70 t
Final weight (2d stage empty; 3d stage full)	$W_{1,II}$	200 t
Mass ratio	E_{II}	4.5
Ratio empty weight 2d stage to propellant weight	k_{II}	0.10
Propellant weight	$W_{P,II}$	700 t
Rate of propellant flow	$\dot{W}_{II}$	5.6 t/sec
Exhaust velocity	u_{II}	2,800 m/sec
Nozzle exit pressure	$p_{e,II}$	24.1 g/cm²
Nozzle exit area	$A_{e,II}$	300 m²
Initial acceleration (absolute)	$a_{0,II}$	17.4 m/sec²
Final acceleration (absolute)	$a_{1,II}$	78.8 m/sec²
Burning time	t_{II}	124 sec
Cut-off altitude	y_{II}	64 km
Cut-off velocity	v_{II}	6,420 m/sec
Cut-off distance (from take-off)	x_{II}	534 km
Angle of elevation at cut-off	φ_{II}	2.5°
Distance between take-off and landing (2d stage)	$x_{L,II}$	1,459 km
Length of 2d stage	L_{II}	14 m
Aft diameter of 2d stage	D_{II}	20 m
Front diameter of 2d stage	D_{III}	9.8 m

TABLE 4. THIRD STAGE (MANEUVER OF ASCENT)

Thrust	F_{III}	200 t
Initial weight	$W_{0,III}$	130 t
Empty weight without payload	$W_{E,III}$	22 t
Final weight after maneuver of ascent	$W_{1,III}$	78.5 t
Ratio empty weight of 3d stage to its total propellant load	k_{III}	0.265
Propellant load (total)	$W_{P,III}$	83 t
Propellant load (maneuver of ascent)	$W_{P,III,01}$	51.5 t
Rate of propellant flow	$\dot{W}_{III}$	702 kg/sec
Exhaust velocity	u_{III}	2,800 m/sec
Nozzle exit pressure	$p_{e,III}$	10 g/cm²
Nozzle exit area	$A_{e,III}$	74 m²
Initial acceleration (absolute)	$a_{0,III}$	15.1 m/sec²
Final acceleration (absolute)	$a_{1,III}$	25.0 m/sec²
Burning time	$t_{III,01}$	73 sec
Cut-off altitude	y_{III}	102 km
Cut-off velocity	v_{III}	8,260 m/sec
Cut-off distance from take-off	x_{III}	1,054 km
Angle of elevation at cut-off	φ_{III}	0°

TABLE 5. THIRD STAGE (MANEUVER OF ADAPTATION)

Thrust	F_{III}	200 t
Initial weight	$W_{1,III}$	78.5 t
Final weight after adaptation maneuver	$W_{2,III}$	66.6 t
Burning time	$t_{III,12}$	17 sec
Velocity increment for adaptation	$v_{III,12}$	460 m/sec

TABLE 6. THIRD STAGE (RETURN MANEUVER IN THE ORBIT)

Thrust (half-throttle)	$\frac{1}{2}F_{III}$	100 t
Initial weight	$W_{3,III}$	32.2 t
Final weight (landing weight)	$W_{4,III}$	27 t
Landing payload	$W_{N,III,34}$	5 t
Burning time	$t_{III,34}$	14.8 sec
Velocity decrement for return maneuver	$v_{III,34}$	480 m/sec
Altitude of perigee of landing ellipse	y_p	80 km
Landing speed	v_L	105 km/h

TABLE 7. THIRD STAGE (DIMENSIONS)

Length of body	L_{III}	15 m
Aft diameter	D_{III}	9.8 m
Wing area	S_W	368 m²
Wing span	b_W	52 m

TABLE 8. TOTAL SHIP (MAIN DATA)

Over-all length without fins	L_{tot}	60 m
Maximum diameter	D_I	20 m
Dry payload	$W^*_{N,III,02}$	25 t
Total payload, including propellant reserves which can be delivered in orbit of departure, after deduction of requirements for return maneuver	$W_{N,III,02}$	39.4 t
Total propellant supply	$W_{P,tot}$	5,583 t

2. THE ORBIT OF DEPARTURE

Let the orbit of departure be such that a satellite vehicle in it will require $T_1 = 7,200$ sec to complete a circle around the earth. The orbital radius will then be $R_{S,E} = 8.11 \cdot 10^8$ cm. The corresponding orbital velocity is $v_{ci,1} = 7.07 \cdot 10^5$ cm sec^{-1} and the orbit must lie in the plane of the ecliptic, while the direction of rotation is the same as that of the earth. Rocket launchings towards the orbit of departure will take place from some point on the equator in order to make maximum use of the circumferential velocity due to the earth's spin.

The plane of the orbit of departure is inclined to that of the equator at an angle $\rho = 23.5°$. The oblateness of the earth causes a slow but continuous retrograde motion of the nodes (or points of intersection of the orbit and the plane of equator) which results in a cyclic change of the angle between orbit of departure and ecliptic between the extreme values of $2 \cdot 23.5 = 47°$ and zero. The period of this nodal regression for the orbit of departure is approximately 90 days. Inasmuch as the instant of departure of the Mars expedition is governed by a certain required position of earth and Mars relative to one another (see section 27), and is thus well known in advance, it is easily possible to situate the nodes of the orbit of departure at the beginning of the assembly operation of the Mars ships in such manner that on the day of departure the orbit will be parallel to the plane of the ecliptic.

We shall see that the total time during which the Mars expedition will be absent from the orbit of departure is 969 days. This is not an even multiple of the period of nodal regression. For this reason any object left behind in the orbit of departure (such as an orbital workshop station) would *not* be circling in the plane of the ecliptic at the time the Mars ships return. The crews returning from Mars must therefore be brought back to the earth's surface by ferry ships ascending in the plane of the actual orbit of return whose orbital radius and period of revolution may still be equal or similar to those of the orbit of departure.

3. TERMINAL VELOCITY OF THIRD STAGE

The third stage will reach its terminal velocity on a track parallel to earth's surface at an altitude of $y_{III} = 0.102 \cdot 10^8$ cm; the selection of this altitude will be accounted for later. The radial distance of the point of thrust cut-off from the geocenter—this cut-off point being simultaneously the perigee of the ellipse of ascent—is then $R_{III} = 6.482 \cdot 10^8$ cm. As shown in Fig. 1, the length of the major semi-axis of the ellipse of ascent follows as

$$a = \tfrac{1}{2}(R_{III} + R_{s,E}) = 7.290 \cdot 10^8 \text{ cm.} \qquad (3.1)$$

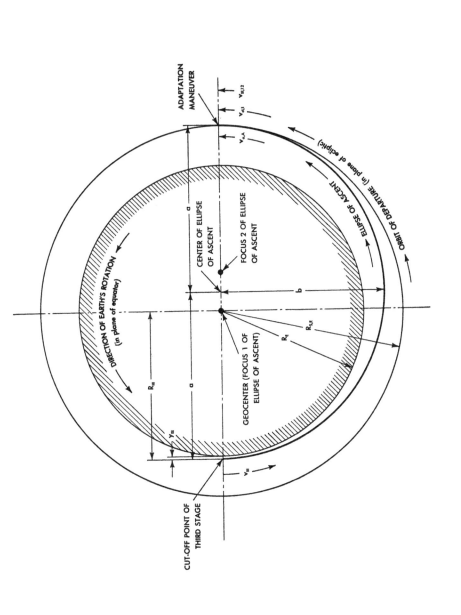

Figure 1. Ellipse of ascent for ferry vessel.

The circular velocity at the altitude of thrust cut-off is:

$$v_{ci,\text{III}} = 7.850 \cdot 10^5 \text{ cm sec}^{-1}.$$

This gives for the perigeal velocity reached by the third stage at thrust cut-off:

$$v_{\text{III}} = v_{ci,\text{III}} \left(2 - \frac{R_{\text{III}}}{a} \right)^{\frac{1}{2}} = 8.260 \cdot 10^5 \text{ cm sec}^{-1}. \quad (3.2)$$

4. POWERED TRACK OF ASCENT

a. First stage

The most suitable ascent track during combustion of first-stage propellants is computed in the light of the following basic considerations:

1. Maximum velocity at cut-off;
2. Vertical launching (take-off);
3. Accelerations normal to any track tangent, as expressed by the equation $n = v\dot\varphi$, shall be limited by structural considerations to $2\,g$.
4. Velocity at cut-off, cut-off altitude, and angle of elevation must be maintained in such relation to one another that decelerations applied to the exhausted first stage by the opening of its parachute do not approach the destructive.

The ascent track is computed by numerical integration after the method of Runge and Kutta. Thus it is first necessary to seek by iteration a track profile which is in accordance with conditions 2, 3, and 4 cited above. The instantaneous acceleration tangential to the track is expressed as follows:

$$\dot v = \frac{F_{\text{I}} + A_{e,\text{I}}\,(p_{e,\text{I}} - p_a) - c_D \cdot A_{\text{I}}\,v^2\gamma/2g_0}{1/g_0\,(W_{0,\text{I}} - \dot W_{\text{I}} \cdot t)} - g_0 \cos\vartheta \quad (4.1)$$

where $g_0 = 981$ cm sec^{-2} represents normal gravitational acceleration, t the instantaneous time of flight, ϑ the trajectory angle to the local vertical, and c_D the drag coefficient of the entire craft. Table 2 contains the other symbols used, together with their numerical values. The craft's frontal area $A_{\text{I}} = 3{,}140 \cdot 10^3$ cm^2 is obtained from the diameter $D_{\text{I}} = 20 \cdot 10^2$ cm of the first stage.

The variation of the drag coefficient c_D dependent upon the Mach number M is shown in Table 9 and was derived by computation of the pressure distribution in consideration of the aerodynamic form of the craft at take-off.

TABLE 9. COEFFICIENT OF DRAG AS A FUNCTION OF MACH NUMBER

M	c_D
>5	0.55
5	0.55
4	0.57
3	0.59
2	0.69
1–1.4	0.80
subsonic	0.40

The ascent track computed with these assumptions has been tabulated in Table 10.

TABLE 10. POWER TRACK OF FIRST STAGE

Time	Acceleration	Velocity	Altitude	Distance	Trajectory angle to the local vertical	
$\Sigma \Delta t$	$\dot{v} \cdot 10^{-2}$	$v \cdot 10^{-2}$	$y \cdot 10^{-2}$	$x \cdot 10^{-2}$	ϑ	
(sec)	(cm sec^{-2})	(cm sec^{-1})	(cm)	(cm)	(deg)	(min)
0	8.8321	0	0	0	0	0
4	9.5164	36.74	72.4	0	0	0
8	10.2724	76.23	297.2	0	0	0
16	12.0320	165.18	1,252.5	21.5	2	58
24	13.9157	269.31	2,971.5	190.5	8	2
32	15.7022	385.94	5,528.9	706.4	14	39
40	19.3283	524.52	8,949.4	1,883.8	23	19
48	25.1416	700.60	13,197.8	4,245.1	34	43
56	32.3347	924.82	18,068.7	8,476.5	46	8
64	40.8963	1,216.72	23,423.9	15,080.9	54	48
72	51.7102	1,585.23	29,260.8	24,574.2	61	25
80	66.5597	2,054.96	35,565.3	37,606.3	66	29
86	82.3034	2,498.78	40,644.7	50,235.6	69	30

Thus the first stage would burn for 86 seconds and reach a velocity of $2.498 \cdot 10^5$ cm sec^{-1} at propellant exhaustion. If actual velocity at cut-off be reduced to $v_I = 2.350 \cdot 10^5$ cm sec^{-1}, residual propellants will amount to $103 \cdot 10^6$ gm corresponding to a thrust period of 1.85 seconds. (It is desirable to provide a propellant reserve of this order of magnitude as a precautionary measure in the event of inaccurate maintenance of the mixture ratio, precluding premature exhaustion of either propellant. Any residual propellants would preferably be jettisoned by the feed pump immediately after the release of the first stage, thus minimizing the mass to be handled by the chute and landing rockets.)

If cut-off takes place at 84 seconds flight time, cut-off altitude will be about $y_I = 40 \cdot 10^5$ cm, horizontal range at cut-off about $x_I = 50 \cdot 10^5$ cm. The trajectory at cut-off will be inclined to the local vertical $\vartheta_I = 69.5°$. The trajectory angle of elevation will consequently be $\varphi_I = 20.5°$ to the local horizontal.

b. Second stage

The parachute landing of the second stage will be favorably affected by the following initial conditions for chute deployment:

Altitude of cut-off: $y_{II} = 64 \cdot 10^5$ cm.

Trajectory angle of elevation at cut-off: $\varphi_{II} = 2.5°$.

Aerodynamic drag is negligible during the thrust period of the second stage, so its velocity increase and horizontal distance may be determined by the use of a closed equation. Its terminal velocity may be expressed with good exactitude by the expression:

$$v_{II}^* = u_{II} \log_e E_{II} - g \sin \varphi_m \frac{W_{0,II}}{\dot{W}_{II}} \frac{E_{II} - 1}{E_{II}} c_r + v_I \quad (4.2)$$

where φ_m represents the mean angle of ascent of the second stage's power track. It is determined by the relation between its altitude increase and horizontal distance as $\varphi_m = 2° \, 51'$. The acceleration of gravity affecting the second stage at mean altitude is $g = 970$ cm sec^{-2}. c_r is the so-called "centrifugal reduction coefficient." The following expression is generally valid:

$$c_r = 1 - \frac{v^2}{g_0 (R_E + y)} . \quad (4.3)$$

For the mean velocity and mean altitude of the second stage we have $c_r = 0.678$.

The data compiled in Table 3 substituted in equation (4.2) yield $v_{II}^* = 6.51 \cdot 10^5$ cm sec^{-1} at complete exhaustion of propellants of the second stage. The horizontal distance traveled by the second stage during the propulsive phase is expressed by

$$\Delta x_{II}^* = \frac{W_{0,II}}{\dot{W}_{II}} \cos \varphi_m \left\{ u_{II} \frac{1}{E_{II}} [E_{II} - 1 - \log_e E_{II}] \right.$$
$$\left. - c_r \frac{g}{2} \sin \varphi_m \frac{W_{0,II}}{\dot{W}_{II}} \left(\frac{E_{II} - 1}{E_{II}} \right)^2 + v_I \frac{E_{II} - 1}{E_{II}} \right\}. \quad (4.4)$$

Substituting identical values, this equation yields $\Delta x_{II}^* = 491 \cdot 10^5$ cm.

Let us, however, utilize an actual terminal velocity at cut-off amounting to only $v_{II} = 6.42 \cdot 10^5$ cm sec^{-1}. This will leave residual propellants amounting to about $6.45 \cdot 10^6$ gm, equivalent to a thrust period of 1.15 seconds, in the second-stage tanks. This shortens the horizontal distance gained to $\Delta x_{II} = 484 \cdot 10^5$ cm.

The total range at second-stage cut-off, as reckoned from the take-off point, is then $x_{II} = x_I + \Delta x_{II} = 534 \cdot 10^5$ cm. Burning time to complete exhaustion of propellants would be

$$t_{II}^* = \frac{W_{0,II} - W_{1,II}}{\dot{W}_{II}} = 125 \text{ sec.}$$

In view of the propellant reserve of 1.15 seconds, actual burning time would be about 124 seconds.

Check on the angle of ascent φ_m:

$$\tan \varphi_m = \frac{y_{II} - y_I}{\Delta x_{II}} = 0.0497; \qquad \varphi_m = 2° \, 51'.$$

c. Third stage

The required terminal velocity of the third stage has been determined in equation (3.2) to be $v_{III} = 8.260 \cdot 10^5$ cm sec^{-1}. Since velocity at cut-off of the second stage is already $v_{II} = 6.42 \cdot 10^5$ cm sec^{-1}, there remain $\Delta v_{III} = v_{III} - v_{II} = 1.84 \cdot 10^5$ cm sec^{-1} to be gained.

A part of this velocity increment is provided by the peripheral velocity v_p of the point of take-off (rotation of the earth). For an equatorial launching site, we have $v_p = 0.464 \cdot 10^5$ cm sec^{-1}. Because the orbit of departure lies in the plane of the ecliptic, however, only a component of this peripheral velocity $v_{rot} = v_p \cos \rho = 0.425 \cdot 10^5$ cm sec^{-1}, corresponding to the angle $\rho = 23.5°$ between the plane of the ecliptic and that of the equator, augments the climb, so the third stage's own thrust must provide a velocity increment $v_{III,01} = \Delta v_{III} - v_{rot} = 1.415 \cdot 10^5$ cm sec^{-1}.

We may neglect the potential energy gained during the climb of the third stage since, at the velocities involved, the mean gravitational effect is approximately balanced by the centrifugal force exerted normal to the trajectory ($c_r \cong 0$). To compute the required mass ratio, we may therefore use

$$\log_e \frac{W_{0,III}}{W_{1,III}} = \frac{v_{III,01}}{u_{III}} = 0.505. \tag{4.5}$$

If the initial weight of the third stage equals $W_{0,III} = 130 \cdot 10^6$ gm, the terminal weight at third-stage cut-off will then be $W_{1,III} = 78.5 \cdot 10^6$ gm. The propellant consumption during thrust period is found as $W_{P,III,01} = 51.5 \cdot 10^6$ gm, being the difference between initial and terminal weights, while the burning time is

$$t_{III,01} = \frac{W_{P,III,01}}{\dot{W}_{III}} = 73 \text{ sec.}$$

5. MANEUVER OF ADAPTATION

Kepler's second law states (see Fig. 1) that

$$R_{III} \cdot v_{III} = R_{S,E} \cdot v_{a,A} \tag{5.1}$$

where $v_{a,A}$ represents the velocity at apogee of the ellipse of ascent. We thus find $v_{a,A} = 6.610 \cdot 10^5$ cm sec^{-1}. However, the circular velocity in the orbit of departure amounts to $v_{ci,1} = 7.07 \cdot 10^5$ cm sec^{-1}; so the velocity of the third stage at apogee must be augmented by $v_{III,12} = v_{ci,1} - v_{a,A} = 0.460 \cdot 10^5$ cm sec^{-1}. We compute the mass ratio for this maneuver as follows:

$$\log_e \frac{W_{1,III}}{W_{2,III}} = \frac{v_{III,12}}{u_{III}} = 0.164. \tag{5.2}$$

Accordingly, the weight of the third stage at the termination of the maneuver of adaptation is $W_{2,III} = 66.6 \cdot 10^6$ gm. The burning time of the adaptation maneuver is

$$t_{III,12} = \frac{W_{1,III} - W_{2,III}}{\dot{W}_{III}} = 17 \text{ sec.}$$

6. COASTING PERIOD IN THE SEMI-ELLIPSE OF ASCENT

Lunar orbital data and Kepler's third law, together with equation (3.1), permit the full circum-tellurian period of revolution in the ellipse of ascent to be computed as follows:

$$T_{ell} = 3.13 \cdot 10^{-10} a^{3/2} = 6,140 \text{ sec.} \tag{6.1}$$

Time on the semi-ellipse is one half of the above, with burning time for the maneuver of adaptation deducted, i.e., $t_{ell} = 3,053$ sec $= 50$ min 53 sec.

7. REQUIRED WING AREA OF THIRD STAGE

The third stage is equipped with wings for the purpose of returning to earth and making a conventional aerodyne landing. Such wings may be integral with the stage itself or they may be retractable. They might be carried disassembled in the cargo space of the third stage and be mounted outside during circling in the orbit of departure. After the payload carried by the third stage has been unloaded in the orbit, and we wish to return to earth, a short deceleration thrust is applied which decreases the orbital velocity of the ship enough to situate the perigee of the landing ellipse at an altitude of 80 kilometers. This brings the ship into an extended earthward coasting path, attended by an increasing gravitational effect throughout the first half of the landing ellipse. Then the wings begin to engage the outer atmospheric layers, at first exerting negative lift, and the vehicle begins to decelerate. After the velocity has been diminished below local circular speed, the lift must become positive and the vessel gradually assumes the characteristics of a relatively conventional aerodyne which approaches earth in an extended decelerated glide. It finally lands conventionally upon a retractable tricycle gear. Thus the design wing area S_W is determined by the acceptable landing speed v_L:

$$S_W = \frac{2g_0 W_{4,III}}{c_{L,max} \gamma_0 v_L^2}. \tag{7.1}$$

In the above formula $c_{L,max} = 1.3$ is the optimum lift coefficient attainable with trailing-edge and leading-edge flaps. $\gamma_0 = 1.293 \cdot 10^{-3}$ gm cm^{-3} is the specific weight of air at ground level. $W_{4,III} = 27 \cdot 10^6$ gm is the weight of the third stage at landing. This gives for a landing speed of $v_L = 2,925$ cm sec$^{-1} = 105$ km/hr a wing area of $S_W = 3.68 \cdot 10^6$ cm^2.

8. NEGATIVE LIFT AT THE PERIGEE OF THE LANDING ELLIPSE

The circular velocity at perigeal altitude $y_p = 8 \cdot 10^6$ cm is $v_{ci,p} = 7.85 \cdot 10^5$ cm sec^{-1}. The acceleration of gravity at this altitude must equal the centrifugal acceleration of an equivalent *free* circular orbit and is $g_p = 955$ cm sec^{-2}. The perigeal velocity of the landing ellipse is found with the help of equation (3.2) and equals $v_p = 8.27 \cdot 10^5$ cm sec^{-1}. The centrifugal acceleration of a *forced* circum-tellurian flight path at perigeal velocity v_p and perigeal altitude y_p is

$$a_p = \frac{v_p^2}{R_E + y_p} = 1{,}052 \text{ cm sec}^{-2}.$$

To force such a flight path will require a centripetal acceleration of $a_z = a_p - g_p = 97$ cm sec^{-2} which must be provided by negative lift of the wings.

The landing weight of the third stage equals $W_{4,III} = 27 \cdot 10^6$ gm requiring a total negative lift of

$$-L = \frac{W_{4,III}}{g_0} a_z = 2.67 \cdot 10^6 \text{ gm.} \tag{8.1}$$

The specific weight of air at altitude $y_p = 8 \cdot 10^6$ cm is $\gamma_p = 2.5 \cdot 10^{-8}$ gm cm^{-3}. In high supersonic regions and for the chosen configuration, lift coefficients of the order of magnitude of $c_L = 0.1$ are easily obtainable without the use of excessive angles of attack. With this lift coefficient the wing area, S_W, as determined in Section 7, can produce a negative lift of

$$-L' = \frac{c_L S_W v_p^2 \gamma_p}{2g_0} = 3.2 \cdot 10^6 \text{ gm.} \tag{8.2}$$

A comparison of equations (8.1) and (8.2) shows that the ship can easily be forced into a circular circum-tellurian flight path at a perigeal altitude $y_p = 8 \cdot 10^6$ cm.

9. RETURN MANEUVER IN THE ORBIT

According to equation (5.1) a perigeal velocity of $v_p = 8.27 \cdot 10^5$ cm sec^{-1} at altitude $y_p = 8 \cdot 10^6$ cm corresponds to an apogeal velocity of

$$v_{a,L} = \frac{R_E + y_p}{R_{S,E}} \cdot v_p = 6.59 \cdot 10^5 \text{ cm sec}^{-1}. \tag{9.1}$$

It is therefore essential to decelerate in the orbit of departure by an amount equivalent to

$$v_{III,34} = v_{ci,1} - v_{a,L} = 0.480 \cdot 10^5 \text{ cm sec}^{-1}. \tag{9.2}$$

The mass ratio required is determined by appropriate use of equation (4.5) and equals 1.19. If the empty weight of the third stage is $W_{E,III} = 22 \cdot 10^6$ gm and a useful load under landing conditions of $W_{N,III,34} = 5 \cdot 10^6$ gm is envisaged, the total weight at departure from the orbit of departure will be, with $W_{4,III} = 27 \cdot 10^6$ gm,

$$W_{3,III} = 1.19 \cdot W_{4,III} = 32.2 \cdot 10^6 \text{ gm.} \tag{9.3}$$

Hence the quantity of propellants needed for the landing maneuver will be $W_{P,III,34} = W_{3,III} - W_{4,III} = 5.2 \cdot 10^6$ gm.

If we now postulate that each ferry vessel carry sufficient propellants for a trip to the orbit and back, that is to say, does not replenish its tanks in the orbit itself, we may compute the disposable load which each vessel can deliver to the orbit. It amounts to

$$W_{N,III,02} = W_{2,III} - W_{3,III} + W_{N,III,34} = 39.4 \cdot 10^6 \text{ gm.} \quad (9.4)$$

Of course, some part of this disposable load will represent propellant reserve for the third stage, although ordinarily such reserves will not be consumed. Mars's ship propellants are identical with those used in ferry vessels, hence excess reserve propellants may be stored in tankage provided in the orbit against future use when the expedition leaves the orbit. Thus there will be no major loss of available transport tonnage. Dry payload may thus be limited to $W^*_{N,III,02} = 25 \cdot 10^6$ gm.

10. AERODYNAMIC ASSUMPTIONS CONCERNING THE WINGED THIRD STAGE

The calculations have been made taking into consideration a lift coefficient $c_L = 0.1$ above Mach 8, with a drag coefficient $c_D = 0.05$, both referring to the wing area of S_W. Table 11 shows the assumptions for lower Mach numbers. The figures correspond to a moderate mean angle of attack during the glide.

TABLE 11. AERODYNAMIC ASSUMPTIONS FOR THIRD STAGE

M	$\epsilon = c_D/c_L$	c_L	c_D
5	0.38	0.144	0.055
3	0.28	0.235	0.065
2	0.24	0.338	0.081
1	0.19	0.343	0.066
subsonic	0.17	0.315	0.053

The figures referred to are based on relatively rough estimates and extrapolations, and represent extremely modest aerodynamic performance caused primarily by the large diameter of the tail section of the third stage. (It should be noted here that a long range glide is in no way essential when descending from the orbit of departure. For this reason thick wing profiles may well be envisaged, thus facilitating the use of retractable wings—despite the fact that thick wing sections are usually considered not applicable to supersonic flight.)

11. GLIDE PATH OF THE THIRD STAGE

Centrifugal force partially supports the weight of the third stage when gliding at higher speeds. Thus the full landing weight $W_{4,III}$ need not be carried by the wings, the portion thereof so carried being $c_r \cdot W_{4,III}$, where c_r is the centrifugal reduction coefficient defined in equation (4.3). The relation between c_r and the velocity v for a mean gliding altitude y is shown in Table 12.

TABLE 12. CENTRIFUGAL REDUCTION COEFFICIENT c_r AS A FUNCTION OF THE VELOCITY FOR AN AVERAGE ALTITUDE $y = 50 \cdot 10^5$cm

$v \cdot 10^{-5}$ (cm sec^{-1})	c_r
8	-0.100
7	0.227
6	0.428
5	0.603
4	0.746
3	0.857
2	0.936
1	0.983

The air density sufficient to sustain the third stage aerodynamically, assisted by the centrifugal reduction coefficient, is

$$\gamma = \frac{W_{4,III}}{S_W} \cdot c_r \cdot \frac{2g_0}{c_L} \cdot \frac{1}{v^2} . \tag{11.1}$$

The altitude y corresponding to such density is obtained by the following atmospheric formula

$$\frac{\gamma}{\gamma_0} = e^{-y/y^*} \tag{11.2}$$

where $y^* = 7.8 \cdot 10^5$ cm represents that altitude at which the density diminishes to the eth part. γ_0 is the density at sea-level: $\gamma_0 = 1.293 \cdot 10^{-3}$ gm cm^{-3}.

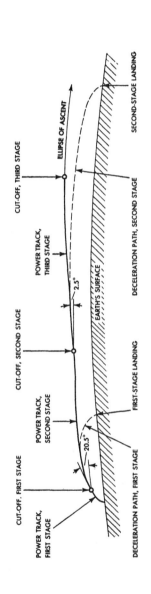

Figure 2. Tracks of powered ascent for the three stages of the ferry vessel and deceleration paths of the first and second stages.

Multiplying the instantaneous lift $L = c_r W_{4,III}$ by ϵ, one derives the instantaneous drag D. The expression

$$-\dot{v} = \frac{D \cdot g_0}{W_{4,III}} = c_r \cdot \epsilon \cdot g_0 \qquad (11.3)$$

then yields the deceleration in flight direction.[1]

The above relations permit determination of the glide path by numerical integration. The results of so doing are shown in Table 13.

TABLE 13. GLIDE PATH OF THIRD STAGE

Velocity $v \cdot 10^{-5}$ (cm sec^{-1})	Altitude $y \cdot 10^{-5}$ (cm)	Deceleration $-\dot{v} \cdot 10^{-2}$ (cm sec^{-2})	Time span Δt (sec)	Distance proceeded $\Delta x \cdot 10^{-5}$ (cm)	Rate of descent $\dot{y} \cdot 10^{-2}$ (cm sec^{-1})	Total time $\Sigma \Delta t$ (sec)	Total distance $\Sigma \Delta x \cdot 10^{-5}$ (cm)
8.27	80						
		0.79	1,610	12,300	4.4	1,610	12,300
7	72.9						
		1.61	622	4,050	10.4	2,232	16,350
6	65.8						
		2.54	394	2,170	13.5	2,626	18,520
5	60.5						
		3.30	303	1,365	15.8	2,929	19,885
4	55.7						
		3.94	254	888	21.2	3,183	20,773
3	50.3						
		4.42	226	565	28.8	3,409	21,338
2	43.8						
		3.43	292	438	16.4	3,701	21,776
1	33.3						
		2.33	300	195	50	4,001	21,971
sonic speed	24.0						

The total time of flight from the perigee of the landing ellipse to sonic speed is thus 4,001 sec = 1 hr 6 min 41 sec. The length of the glide path involved is 21,971 km, or about 55 per cent of the earth's circumference. The glide is subsonic from the altitude of 24 km down.

[1] The relation $-\dot{v} = c_r \epsilon g_0$ is only applicable after the lift has become *positive*. c_r is negative at and immediately after entry at the perigee of the landing ellipse. At a decreasing rate it approaches and passes zero, then becoming positive. Accordingly, the glide will begin with negative lift; zero lift will coincide with the local circular velocity, after which positive lift will gradually increase. The longitudinal negative acceleration of the ship $-\dot{v}$ acts throughout the atmospheric glide and its sign does *not* change.

Coasting flight through the landing semi-ellipse lasts about 51 minutes and makes another half-circle of the earth. Thus the whole return from the orbit of departure to touch-down requires about two hours, requiring the third stage to practically circle the earth.

12. SKIN TEMPERATURE RISE DURING GLIDE OF THE THIRD STAGE

A body exposed to a high velocity air flow attains a temperature higher than the temperature T_a of the ambient air. Once the body has reached a stationary temperature at which no radiation losses take place and when there is no inward heat flow, it is said to have the so-called "natural temperature" T_n. This is somewhat lower than the adiabatic stagnation temperature T_{st}. In the case of cones or wedges pointing into the air flow and having cone or wedge angles of between 15 and 50 degrees, the relation between T_n and T_{st} according to Busemann is

$$T_n - T_a = 0.89 \, (T_{st} - T_a). \tag{12.1}$$

The relation between the difference $T_n - T_a$ and velocity of flight v is shown in Table 14 as computed with the use of enthalpy tables. No dissociation of air molecules was here considered, there being reason to believe that it exerts but little effect on heat transfer under conditions here postulated (re-association along the relatively cool skin).

TABLE 14. $T_n - T_a$ **AS A FUNCTION OF VELOCITY** v

$v \cdot 10^{-5}$ (cm sec^{-1})	$T_n - T_a$ (°K)
8.27	22,673
7	16,256
6	12,518
5	8,995
4	5,548
3	3,161
2	1,581
1	442

The altitudes y at which the above velocities occur during the downward glide of the third stage have already been compiled in Table 13. The air temperatures at these altitudes have been measured by instrumentation in high-altitude rockets. In this manner we obtain air temperatures T_a and natural temperatures T_n for a series of typical flight conditions of the third stage. They are tabulated in columns 3 and 4 of Table 15.

Heat absorption from the boundary layer into the skin per unit of area is in accordance with

$$Q_{in} = h\,(T_n - T_s) \quad [\text{cal cm}^{-2}\,\text{sec}^{-1}] \qquad (12.2)$$

where T_s is the instantaneous skin temperature and h is the coefficient of heat transfer. Eber's equation for the heat transfer coefficient

$$h = (0.0071 + 0.0154\beta^{0.5})\,\frac{1}{l^{0.2}}\left(\frac{\gamma}{g_0}\right)^{0.8}\frac{k}{\mu^{0.8}}\,v^{0.8} \qquad (12.3)$$

has proved reliable. It is applicable in cases of turbulent boundary layers around conical or wedge-shaped bodies and is valid in regions of moderate supersonic speeds. β represents the cone or wedge angle of the nose or leading edge, l the down-stream length from the tip of the cone or leading edge to the point of interest, γ the local specific gravity of the ambient air, k the heat conductivity of air, and μ the dynamic viscosity of air.

The local value of γ may be obtained from equation (11.2). Let us arbitrarily select a leading-edge angularity of $\beta = 20°$ and a distance therefrom of $l = 32$ cm. The variation of the heat transfer coefficient h derived therefrom, as taking place during the earthward glide of the third stage, is tabulated in column 5 of Table 15.

Table 13 reveals that the third stage will remain for extended periods under each of the transient flight conditions. We may therefore neglect the heat capacities of the thin fuselage and wing skinning and assume that thermal balance is continually attained. In such a thermal balance, the heat transferred to the skin Q_{in} will be currently dissipated by radiation loss Q_{out}:

$$Q_{in} = Q_{out}. \qquad (12.4)$$

(Heat loss to the interior of the craft is herein neglected. Such heat loss to the interior must be held within as small limits as possible to minimize the demands on the structural members of the fuselage and the cabin cooling system.)

According to the Stefan-Boltzmann law, skin radiation is computed by

$$Q_{out} = 1.38 \cdot 10^{-12} \cdot \epsilon \cdot T_s^4 \quad [\text{cal cm}^{-2} \text{ sec}^{-1}]. \quad (12.5)$$

The emissivity for sheet steel is $\epsilon = 0.96$. Applying equations (12.2), (12.4), and (12.5) we obtain

$$h\,(T_n - T_s) = 1.32 \cdot 10^{-12}\,T_s^4. \quad (12.6)$$

By substituting the numerical values of T_n and h (columns 4 and 5 of Table 15), this relation will reveal the temperature T_s at which heat balance is attained for the transient flight conditions through which the third stage passes during the glide. Results are tabulated in column 6 of Table 15.

TABLE 15. SKIN HEATING DURING GLIDE OF THIRD STAGE

$v \cdot 10^{-5}$ (cm sec^{-1})	$y \cdot 10^{-5}$ (cm)	T_a (°K)	T_n (°K)	h (cal cm^{-2} sec^{-1} °K^{-1})	T_s (°K)
8.27	80	200	22,873	0.456	933
7	72.9	217	16,473	0.795	983
6	65.8	270	12,788	1.069	989
5	60.5	300	9,295	1.627	1,005
4	55.7	320	5,868	2.139	945
3	50.3	322	3,483	2.682	855
2	43.8	300	1,881	3.790	754
1	33.3	225	667	5.692	510

As shown in this table, the skin attains a maximum temperature of 1,005° K, or 732° C. It would seem just within acceptable limits for scale-resistant steel alloys of good tensile strength at elevated temperatures. Warping of the fuselage and distortion of wing contours could be prevented by application of a shingle-like skin structure.

Equation (11.1) shows that a lighter wing loading $W_{4,\text{III}}/S_W$ leads to the third stage being capable of aerodynamic sustention at a lower density γ at identical speeds, which is to say, a higher altitude. A lower density γ, according to equation (12.3) produces a lower heat transfer coefficient and hence leads to a lower temperature at which the skin attains temperature equilibrium. Thus lowering the wing loading, i.e., increasing the wing area, will lower skin temperatures if this be thought desirable.

Equation (12.3) has not yet been proved at extremely high Mach numbers, which is limiting to a degree. Recently several authors have suggested modifications to equation (12.3). With the use of these modifications skin temperatures some 300° C higher than those here tabulated have been computed. In the light of the above, it may be essential to utilize lower wing loadings.

13. ROCKET NOZZLE EXIT AREAS

The caliber of the hull is determined by the nozzle exit area in the case of large rocket ships, if a high adiabatic expansion ratio is to be utilized. It is calculated from the throat area of the discharge nozzles and the expansion ratio used.

The throat area for a ratio $\kappa = c_p/c_v = 1.2$ of specific heats is obtained as follows:

$$A_{th} = \frac{\dot{W}}{2.09 \, p_i} \sqrt{RT_i} \tag{13.1}$$

where $\dot{W}$ is the propellant consumption per second, $R = \dfrac{848}{\mu}$ the gas constant, T_i the temperature of combustion, and p_i the combustion chamber pressure.

For combustion of hydrazine and nitric acid in stoichiometric ratio

$$1.25N_2H_4 + HNO_3 \rightarrow 3H_2O + 1.75N_2 \tag{13.2}$$

a combustion temperature of $T_i = 2,850°$ K is found, while the mean molecular weight of the combustion products is $\mu = 21.7$. Let the combustion pressure of all of the three stages be $p_i = 15 \cdot 10^3$ gm cm^{-2}. Using equation (13.1) the values tabulated in column 2 of Table 16 apply to the throat areas of all three stages.

The relation

$$\frac{A_{th}}{A_e} = \left(\frac{\kappa+1}{2}\right)^{\frac{1}{\kappa-1}} \left(\frac{p_e}{p_i}\right)^{\frac{1}{\kappa}} \sqrt{\frac{\kappa+1}{\kappa-1}\left[1 - \left(\frac{p_e}{p_i}\right)^{\frac{\kappa-1}{\kappa}}\right]} \tag{13.3}$$

applies to the ratio between throat nozzle area A_{th} and nozzle exit area A_e. p_e represents the pressure to which gases in the exhaust nozzle are to be expanded.

a. First stage

The rocket motors of the first stage ignite at an ambient pressure of $p_a = 1,000$ gm cm^{-2} (sea level). If a certain degree of hyperexpansion is envisaged, say to $p_{e,\mathrm{I}} = 700$ gm cm^{-2} (which has been shown to be possible), then the use of equation (13.3) shows $A_{th,\mathrm{I}}/A_{e,\mathrm{I}} = 0.265$; and using the values for $A_{th,\mathrm{I}}$ computed according to equation (13.1), we find $A_{e,\mathrm{I}} = 2,240 \cdot 10^3$ cm^2. This would correspond to a circular area having a diameter of 1,690 cm. But enough space must remain in the aft cross-section of the first stage to provide stowage for the large retarding parachute. It is assumed that the parachute container will be inserted in the central axis of the first stage, sandwiched between the rocket motors, and that the chute itself will be ejected rearward. If the chute container's diameter is envisaged as 1,070 cm, the diameter of the first stage as a whole will be 2,000 cm.

b. Second stage

The rocket motor of the second stage comes into use at altitude $y_\mathrm{I} = 40 \cdot 10^5$ cm. The atmospheric pressure at this height is only about 4 gm cm^{-2}. Let us assume the rearmost diameters of the first and second stages to be equal for aerodynamic and design purposes. Then, if a circle of 422 cm diameter be cut from the aft cross-section of the second stage to house the much smaller chute container required, we may compute that the available area for the discharge nozzles is $A_{e,\mathrm{II}} = 3,000 \cdot 10^3$ cm^2. The resulting area ratio $A_{th,\mathrm{II}}/A_{e,\mathrm{II}} = 0.01995$ used in equation (13.3) gives an expansion ratio of $p_i/p_{e,\mathrm{II}} = 622$. With $p_i = 15 \cdot 10^3$ gm cm^{-2} the nozzle exit pressure becomes $p_{e,\mathrm{II}} = 24.1$ gm cm^{-2} which is still a multiple of the atmospheric pressure at the altitude of ignition of the second stage.

c. Third stage

The third stage's thrust does not exert itself until the altitude $y_\mathrm{II} = 64 \cdot 10^5$ cm has been reached where the atmospheric pressure is only about .2 gm cm^{-2}. The outflowing gas cannot be expanded down to any such low pressure, for it would require outsize nozzle discharge areas. Since the third stage must be aerodynamically acceptable (which is not to say highly refined) in view of its subsequent glide to earth, the aerodynamic configuration of the third stage determines the nozzle discharge sections which can be used. Considering a nozzle discharge pressure of $p_{e,\mathrm{III}} = 10$ gm cm^{-2}, equation (13.3) solves for a nozzle area ratio of $A_{th,\mathrm{III}}/A_{e,\mathrm{III}} = 0.0101$, and a nozzle exit area of $A_{e,\mathrm{III}} = 740 \cdot 10^3$ cm^2. No parachute being required for the third stage, its rearmost diameter is computed as about $D_\mathrm{III} = 980$ cm.

Throat and discharge sections of the three stages are compiled in Table 16 for convenient reference.

TABLE 16. NOZZLE THROAT AND EXIT AREAS

Stage	$A_{th} \cdot 10^{-3}$ (cm²)	$A_e \cdot 10^{-3}$ (cm²)
1	595	2,240
2	59.8	3,000
3	7.5	740

14. DIMENSIONS

The determining factors as to dimensions of the three stages are the quantities of propellants required. The weight ratio of the two propellants according to reaction (13.2) is $\dot{W}_{NA}/\dot{W}_{Hy} = 1.57$. The specific gravities of the two chemicals are $\sigma_{NA} = 1.58$ gm cm⁻³ and $\sigma_{Hy} = 1.011$ gm cm⁻³. Thus the ratio of the separate tankages is conveniently $V_{NA}/V_{Hy} = 1$. The mean mixture density, that is to say, the ratio of the entire propellant weight to the entire propellant volume, becomes

$$\sigma_m = 1.295 \text{ gm cm}^{-3}. \tag{14.1}$$

a. First stage

The diameter of the first stage was set at $D_I = 2,000$ cm in the preceding section. The weight of propellants amounts to $W_{P,I} = 4.8 \cdot 10^9$ gm and this, using equation (14.1), corresponds to a tankage volume of $3.7 \cdot 10^9$ cm³. If the inside tank diameter is taken as 1,950 cm, an appropriate cylindrical tank will be 1,240 cm long. The two tanks will be superimposed and, allowing for domed heads with some free space between the tanks, let us add 260 cm. Let the rocket motors have a length of 600 cm, the pump and valve section 500 cm. Atop the first stage there will be mounted a jet deflector requiring 300 cm of length. It will deflect the exhaust gases when ignition of the second stage takes place. Thus the main dimensions of the first stage are set as $D_I = 2,000$ cm, $L_I = 2,900$ cm.

b. Second stage

The rear diameter of the second stage is also $D_{II} = 2,000$ cm. The forward diameter will equal the rear diameter of the third stage, i.e., $D_{III} = 980$ cm. The shape of the second stage is therefore necked-in. The volumetric content of the propellant tanks is $W_{P,II}/\sigma_m = 5.4 \cdot 10^8$ cm³, from which is computed the length of the tank section, namely 500 cm. Taking 450 cm for the length of the rocket motors, 250 cm for the pump and valve section, and 200 cm for the jet deflector against the exhaust gases of the third stage, the length of the second stage will be $L_{II} = 1,400$ cm.

c. Third stage

The fuselage or hull of the third stage calls for the configuration of a stubby artillery shell. The base diameter will be $D_{III} = 980$ cm and a suitable length is $L_{III} = 1,500$ cm. The propellant tanks require a volume of $W_{P,III}/\sigma_m = 6.4 \cdot 10^7$ cm³ and this will call for a length of but 150 cm. If the length of the rocket motors be 300 cm and that of the pump and valve section 200 cm, there remains, just ahead of the propellant tanks, a conical space some 850 cm high and 750 cm across the base. This is available for personnel, cargo, instruments, and other useful load. The required wing area (see section 7) is $S_{III} = 3.68 \cdot 10^6$ cm². One possible conformation is a swept-back, all wing design, with the fuselage in its central axis. If the chord length of the wings at the fuselage is 80 per cent of the length of the fuselage, and if the wing plan is trapezoidal, the span works out at $b_W = 5,200$ cm.

15. WEIGHT ESTIMATE

Table 17 contains more detailed data on the weights of the three stages. These data were based partly on experience with components of smaller liquid rockets and aircraft and partly on relatively crude over-all stress computations. The weights given should be capable of considerable reduction after careful, detailed stress and design studies. This applies particularly to the first stage. Since there exist no experience factors to draw on, applicable to such a large unit, an extremely ample safety factor was applied to it.

All three stages have had the supply of hydrogen peroxide for turbopump actuation added to their empty weight estimates, although this will be consumed during thrust periods, thus gradually diminishing the total weight of the vessel. While the peroxide turbine exhaust produces a very considerable thrust, particularly at greater heights, this has not been included in thrust calculations.

TABLE 17. SUBASSEMBLY WEIGHTS OF THE THREE STAGES (METRIC TONS)

Stage	1	2	3
Fuselage, including tanks	85	14	2.6
Rocket motors	130	16	3.0
Turbopumps	79	8	1.0
Hydrogen peroxide	82	12	1.4
Steam plant, empty	7	1	0.1
Tubing and valves	12	1	0.2
Jet-deflector for next-upper stage	15	2.5	...
Guide and disconnect mechanism for next-upper stage	20	1.5	...
Fins and rudders	40	....	0.5
Retardation chute	85	3.8	...
Chute container and explusion device	15	0.8	...
Brake rockets	40	4	...
Hinged motors and servos	40	3.8	1.0
Flywheels for spatial attitude control	..	....	0.6
Electric power supply	10	1	0.3
Wings	..	....	6.0
Landing gear	..	....	0.7
Pressurized crew spaces, including equipment	..	....	1.9
Airlock and windows	..	....	0.6
Heat insulation and refrigeration plant	..	....	0.3
Air recuperation	..	....	0.4
Automatic guidance, instrumentation and radio equipment	..	....	1.2
TOTAL empty weight	660	69.4	21.8
Empty weight used in performance calculations	700	70	22

A simple comparison between tank volumes and empty weights indicates that all three stages are buoyant in water when empty. This is particularly important for the first and second stages, for it permits them to be salvaged and re-used when landed at sea.

16. LANDING THE FIRST AND SECOND STAGES

a. Retardation chutes

It is desirable to permit relatively high terminal rates of descent in the interest of utilizing chutes of acceptably small dimensions and to reduce the rate of descent to practically zero by downward-firing powder rockets immediately before the stage makes contact with the water. Comparative calculations likewise indicate that this mode of applying retarding forces offers weight-saving advantages.

If the design terminal rate of descent of the first and second stages is $v_z = 5 \cdot 10^3$ cm sec^{-1}, the required retardation area for the first stage will be

$$A_I = \frac{2 \cdot g_0 \cdot W_{E,I}}{c_D \cdot \gamma_0 \cdot v_z^2} = 3.26 \cdot 10^7 \text{ cm}^2 \qquad (16.1)$$

corresponding to a chute canopy diameter of $D_{c,I} = 6.45 \cdot 10^3$ cm. The subsonic drag coefficient of the chute is herein assumed to be $c_D = 1.3$.

Similarly computed, the chute canopy area for the second stage is $A_{II} = 3.26 \cdot 10^6$ cm^2 and it will have a diameter of $D_{c,II} = 2.04 \cdot 10^3$ cm.

The paths of retardation require load factors for the canopies as follows: maximum load factor for first-stage parachute $N_{max} = 6.03$ g; maximum load factor for second-stage parachute $N_{max} = 8.1$ g. (See Tables 18 and 19.)

The chute canopies would be designed along the lines of Madelung's ribbon parachutes, with the difference, however, that the ribbons would consist of thickly woven wire. The shroud lines would be steel cables. A check computation shows that canopy temperatures in both cases would remain acceptable. For the load factors referred to, the computed weights of the retardation chutes for the two stages are respectively 85 metric tons and 3.8 tons. The rather surprising difference in weight between the heavy first-stage chute and the light second-stage chute is accounted for mainly by the greater length of the first stage's canopy shroud lines. In chutes of such dimensions as those under consideration, and in the light of such high loading factors, the risers represent a very high proportion of the total weight of the chutes. The difference in the sizes of the chutes also greatly affects the stowage spaces required for them. (See section 13.)

b. Pre-landing retardation by powder rockets

It is desirable to have the burning periods of the retardation powder rockets relatively short to avoid their performing useless work against gravity. Their required thrust is

$$F_r = \frac{W_E}{g_0} (a_r + g_0). \qquad (16.2)$$

If the burning period of the powder rockets be set at $t_r = 2$ sec, the absolute deceleration is $a_r = v_z/t_r = 25 \cdot 10^2$ cm sec^{-2} and the retardation path $a_r t_r^2/2 = 50 \cdot 10^2$ cm. Thus if the retardation powder rockets are fired at an altitude of 50 meters, say by proximity fuses, the stage will have a vertical velocity of zero at impact. Therefore, should the stage descend at sea, it may be salvaged intact.

The thrust requirement of the retardation rockets of the first stage is computed by the use of equation (16.2) as $F_{r,I} = 24.8 \cdot 10^8$ gm, that of the second stage as $F_{r,II} = 24.8 \cdot 10^7$ gm. If the powder retardation rockets are credited with an exhaust velocity of $u_r = 2 \cdot 10^5$ cm sec^{-1} a calculation involving the law of conservation of momentum indicates that the first-stage retardation rockets will require $W_{P,I} = 24.36 \cdot 10^6$ gm of powder, the second stage $W_{P,II} = 24.36 \cdot 10^5$ gm. As a precautionary measure it is wise to assume that the steel containers of the powder rockets will weigh 65 per cent of their powder charges. Thus the total weight of the retardation rockets of the first stage will be $W_{r,I} = 40 \cdot 10^6$ gm and that of the second stage $W_{r,II} = 4 \cdot 10^6$ gm.

The weight of the full retardation equipment (chute and retardation rockets) of the first stage will amount to 18 per cent of the stage's empty weight, that of the equipment pertaining to the second stage 11 per cent of its empty weight.

c. Paths of retardation

The paths of retardation of the two stages are determined by numerical integration. The retardation along the horizontal component x is

$$\ddot{x} = \frac{c_D \cdot A}{2 W_E} \cdot \gamma \cdot v^2 \cdot \frac{v_x}{\sqrt{v_x^2 + v_y^2}}. \qquad (16.3)$$

The retardation along the vertical component y is

$$\ddot{y} = \frac{c_D \cdot A}{2 \cdot W_E} \cdot \gamma \cdot v^2 \cdot \frac{v_y}{\sqrt{v_x^2 + v_y^2}} \pm g_0 c_r \qquad (16.4)$$

wherein c_r again represents the centrifugal reduction coefficient according to equation (4.3). A supersonic drag coefficient of $c_D = 1.4$ has been assumed. The expression

$$N = \frac{c_D \cdot A \cdot \gamma \cdot v^2}{2 \cdot W_E \cdot g_0} \qquad (16.5)$$

provides the load factor to which the chute is subjected during the retardation process.

Differential equations (16.3) and (16.4) were numerically integrated according to the Runge-Kutta method. The variation of the retardation paths thus computed is tabulated in Tables 18 and 19.

TABLE 18. DECELERATION PATH OF FIRST STAGE

Initial conditions: cut-off velocity $v_I = 2.35 \cdot 10^5$ cm sec^{-1}
  cut-off altitude $y_I = 4 \cdot 10^6$ cm
  angle of elevation at cut-off $\varphi_I = 20.5°$

Total time $\Sigma \Delta t$ (sec)	Velocity $v \cdot 10^{-2}$ (cm sec^{-1})	Altitude $y \cdot 10^{-5}$ (cm)	Angle of elevation φ (deg)	(min)	Load factor N (g)	Total distance $x \cdot 10^{-5}$ (cm)	Remarks
10	2,350	40.60	+20	30	6.03	20.28	$N_{max} = 6.03g$
20	1,994	47.69	18	3	1.87	38.56	
30	1,837	53.17	15	17	0.90	55.82	
40	1,741	57.42	12	16	0.61	72.58	
50	1,675	60.58	9	5	0.42	88.94	
60	1,626	62.72	5	46	0.32	104.97	
70	1,591	63.85	2	20	0.27	120.74	
90	1,564	64.02	−1	10	0.26	151.50	$y_{max} = 64.02$
110	1,524	61.49	8	16	0.31	180.99	$\cdot 10^5$ cm
130	1,484	55.42	15	26	0.52	208.42	
150	1,397	45.97	22	39	1.10	231.34	
160	1,095	34.76	30	23	3.13	239.27	
170	793	29.70	35	27	3.99	244.28	
180	504	25.71	42	17	3.02	247.10	
190	324	22.71	51	35	2.01	248.61	
200	232	20.38	62	28	1.48	249.00	

This computation, if continued, indicates that the first stage will make ground or water contact at sea level after an elapsed period of $\Sigma \Delta t = 400$ sec = 6 min 40 sec from take-off and at a range from the take-off point of $x_{L,I} = 304$ km.

TABLE 19. DECELERATION PATH OF SECOND STAGE

Initial conditions: cut-off velocity $v_{II} = 6.42 \cdot 10^5$ cm sec^{-1}
cut-off altitude $y_{II} = 6.4 \cdot 10^6$ cm
angle of elevation at cut-off $\varphi_{II} = 2.5°$

Total time $\Sigma \Delta t$ (sec)	Velocity $v \cdot 10^{-2}$ (cm sec^{-1})	Altitude $y \cdot 10^{-5}$ (cm)	Angle of elevation φ (deg)	(min)	Load factor N (g)	Total distance $x \cdot 10^{-5}$ (cm)	Remarks
5	6,420	64.0	+2	30	2.74	31.8	
10	6,289	65.3	2	25	2.23	63.0	
15	6,183	66.4	2	16	1.85	93.4	
20	6,086	67.5	2	5	1.59	123.4	
30	6,008	68.5	1	55	1.19	182.3	
40	5,891	70.2	1	37	0.97	240.2	
50	5,796	71.4	1	11	0.83	297.3	
60	5,714	72.1	0	43	0.78	353.7	
70	5,638	72.3	0	0	0.76	409.3	$y_{max} = 72.3$
80	5,565	72.0	−0	16	0.78	464.2	$\cdot 10^5$ cm
90	5,489	71.2	0	47	0.88	518.2	
100	5,402	69.9	1	20	1.07	571.2	
110	5,300	68.1	1	55	1.40	622.8	
120	5,173	65.8	2	31	1.86	672.5	
130	4,984	63.1	3	11	2.48	719.8	
140	4,745	59.9	3	55	3.58	763.6	
150	4,339	56.2	4	47	5.12	802.0	
160	3,861	52.2	5	56	6.93	833.6	
170	3,194	48.1	7	28	8.10	857.3	$N_{max} = 8.1\ g$
180	2,402	44.1	9	40	7.86	873.4	
190	1,642	40.4	13	7	6.02	883.7	
200	1,080	37.1	17	57	3.88	890.4	
210	731	34.1	24	20	2.65	894.7	

This computation, if continued, indicates that the second stage will make ground or water contact at sea level after an elapsed period of $\Sigma \Delta t = 480$ sec $= 8$ min and at a great circle range from the take-off point of $x_{L,II} = 1,459$ km.

Figure 2 gives a scale picture of the total powered ascent track of the three-stage ferry vessel, including the descent paths of the detached stages.

17. ASTRONOMIC AND PHYSICAL DATA APPLYING TO A VOYAGE TO MARS

TABLE 20

		Earth		Mars
Mean distance from sun	a_E	$1.495 \cdot 10^{13}$ cm	a_M	$2.28 \cdot 10^{13}$ cm
Distance from sun at perihelion	$a_{E,min}$	$1.47 \cdot 10^{13}$ cm	$a_{M,min}$	$2.06 \cdot 10^{13}$ cm
Distance from sun at aphelion	$a_{E,max}$	$1.52 \cdot 10^{13}$ cm	$a_{M,max}$	$2.49 \cdot 10^{13}$ cm
Period of revolution	T_E	365 days	T_M	687 days
Mean orbital velocity	v_E	$29.80 \cdot 10^5$ cm sec^{-1}	v_M	$24.10 \cdot 10^5$ cm sec^{-1}
Radius of planet	R_E	$6.380 \cdot 10^8$ cm	R_M	$3.390 \cdot 10^8$ cm
Parabolic escape velocity at surface	$v_{p,E}$	$11.18 \cdot 10^5$ cm sec^{-1}	$v_{p,M}$	$5.04 \cdot 10^5$ cm sec^{-1}
Circular velocity at surface	$v_{ci,E}$	$7.90 \cdot 10^5$ cm sec^{-1}	$v_{ci,M}$	$3.56 \cdot 10^5$ cm sec^{-1}
Radius of selected satellitic orbit	$R_{S,E}$	$8.110 \cdot 10^8$ cm	$R_{S,M}$	$4.390 \cdot 10^8$ cm
Circular velocity of this satellitic orbit	$v_{ci,1}$	$7.07 \cdot 10^5$ cm sec^{-1}	$v_{ci,2}$	$3.14 \cdot 10^5$ cm sec^{-1}
Period of revolution of satellitic orbit	T_1	2 hr 0 min 0 sec	T_2	2 hr 26 min 24 sec
Gravitational acceleration at surface	g_0	1 g_0	g_M	0.38 g_0
Solar constant	. . .	0.1325 watt/cm²	. . .	0.0568 watt/cm²
Albedo	. . .	. . . . . .	. . .	0.27
Atmospheric pressure at surface*	$p_{0,E}$	760 mm Hg	$p_{0,M}$	64 mm Hg
Atmospheric density at surface*	$\gamma_{0,E}$	$1.293 \cdot 10^{-3}$ gm cm^{-3}	$\gamma_{0,M}$	$0.1080 \cdot 10^{-3}$ gm cm^{-3}
Boiling point of water at surface*	. . .	100° C	. . .	44° C
Mean yearly temperature	. . .	16° C	. . .	9° C
Altitude increment at which atmospheric pressure decreases by a factor of 10	. . .	18 km	. . .	47 km

* With respect to Mars, these figures are derived from indirect conclusions based upon measurements available at present. They require further confirmation.

18. DESIGN

The space vessels envisaged for the interplanetary voyage to Mars will be required to perform four main thrust maneuvers. *Between these four main maneuvers, it may be necessary* that minor corrective thrusts be applied.

Maneuver 1: Departure from the "orbit of departure" around the earth. This maneuver induces the ships into a free coast through an "escape hyperbola" which, provided the moment of departure is correctly chosen, leads directly into a circum-solar ellipse whose remotest point from the sun ("aphelion") lies a few thousand kilometers this side of the Martian orbit.

Maneuver 2: Entrance into a satellite orbit around Mars. In this maneuver the ships will be allowed to fall toward Mars through an "approach hyperbola." Shortly before passing through the vertex of this hyperbola, the ships will be decelerated by a rocket counterblast to convert the hyperbolic track into a circular one. The ships can then remain in the ensuing circum-Martian orbit until time of departure to earth.

Maneuver 3: Departure from circum-Martian orbit. This maneuver induces the ships into an "escape hyperbola" which leads directly into the circum-solar ellipse whose closest point to the sun ("perihelion") lies some thousand kilometers outside the earth ("tellurian") orbit.

Maneuver 4: Return into the former "orbit of departure" around the earth. In the vertex of the "approach hyperbola" through which the ships fall in a tangential loop toward the earth, the velocity is retarded to an extent adequate to convert the hyperbolic fall into a circular orbit.

Thus, the space ships are exclusively used between satellite orbits around earth and Mars. This fact makes possible the following deviations from the design of the ferry vessels:

(a) The rocket thrust may become smaller than the ship's weight, since in the satellite orbits the ship's weight is constantly sustained by centrifugal forces. One can get by with surprisingly small power plants which will, however, be operated for relatively long periods during the initial maneuvers when large masses are still involved. For practical reasons (shipping to the orbit of departure) it has been assumed that the space ships are equipped with the same type of rocket power plants as used in the third stage of the ferry vessels. The thrust of this unit is 200 metric tons.

(b) The interplanetary space ships will always operate in a vacuum, which permits us to neglect all forms of streamlining. Nothing even remotely resembling a hull is required. Propellant tankage will be supported in light dural framing. Although tankage volume, particularly for the first maneuver, is very large, stresses are never very high because the unvarying thrust of 200 tons cannot accelerate the ships rapidly when heavy. Therefore, all tanks can be of very light construction.

(c) The multi-stage principle, so essential for the ferry vessels, can be abandoned. One set of propellant containers is provided for each of the four main maneuvers, and these tanks together with their supporting dural structure are jettisoned after completion of the corresponding maneuvers. Any propellants left over after power maneuvers are completed can be pumped into a special set of reserve containers before the tanks are jettisoned. All maneuvers are carried out with the same power plant.

(d) The tanks, as also the crew spaces or "nacelles," will be made of thin-walled, nylon-reinforced plastic. Aside from the advantage of a very considerable weight-saving, this design method permits us to freight the bulky tanks and nacelles up to the orbit of departure in a collapsed condition. This is of vital importance in view of the limited cargo space available in the ferry vessels.

19. "PASSENGER" AND "CARGO" SPACE VESSELS

It is contemplated that only seven of the ten vessels comprising the Mars flotilla shall return to the circum-tellurian orbit of departure. These we shall designate as "passenger vessels." Their primary purpose is the transportation of human beings. The three remaining "cargo vessels" will serve to carry to the circum-Martian orbit the three "landing boats" whose purpose is to permit descent to Mars from the circum-Martian orbit and re-ascent to the same.

The weight of each vessel, whether "passenger" or "cargo," will be identical at departure from the circum-tellurian orbit of departure. This is desirable in order that all vessels may possess identical acceleration characteristics so that their power tracks may likewise be identical in the interest of close stationkeeping.

When the expedition departs from the circum-Martian orbit on its return trip to the circum-tellurian orbit, no "cargo vessel" nor "landing boat" will accompany it. One boat will remain on Mars itself, while the remaining two boats and the three cargo vessels will be abandoned to circle indefinitely in the circum-Martian orbit. Thus, no cargo vessel will be required to perform thrust maneuvers 3 and 4. As a consequence, the load capacity which, in the passenger vessels, must be utilized to carry propellants for maneuvers 3 and 4, is available for payload in the cargo vessels. We shall see that the total payload of any cargo vessel is considerably greater than the full weight of a landing boat. The result is that it is within the capacity of the cargo vessels not only to carry the landing boats, but also sufficient food, water, and oxygen to subsist the entire expedition until departure from the circum-Martian orbit.

20. TABULATION OF MAIN DATA WITH SYMBOL MEANINGS

TABLE 21

Thrust	F	200 t
Total propellant supply		
passenger vessels	$W_{P,P}$	3,662.5 t
cargo vessels	$W_{P,C}$	3,306 t
Rate of propellant flow	$\dot{W}$	702 kg/sec
Exhaust velocity	u	2,800 m/sec
Nozzle exit pressure	p_e	10 gm cm^{-2}
Nozzle exit area	A_e	74 m^2
Total length		
passenger vessels	L_P	41 m
cargo vessels (including landing boats)	L_C	64 m
Maximum diameter (passenger vessels)	D_P	29 m
Maneuver 1		
Initial weight	$W_{0,1}$	3,720 t
Propellant supply (including 3.5% velocity reserve)	$W_{P,1}$	2,814 t
Final weight	$W_{1,1}$	906 t
Initial acceleration	$a_{0,1}$	0.054 g
Final acceleration	$a_{1,1}$	0.214 g
Burning time	t_1	3,965 sec

TABLE 21 (Concluded)

Maneuver 2		
Initial weight	$W_{0,2}$	902 t
Propellant supply (including 10% velocity reserve)	$W_{P,2}$	492 t
Final weight	$W_{1,2}$	410 t
Initial acceleration	$a_{0,2}$	0.22 g
Final acceleration	$a_{1,2}$	0.45 g
Burning time	t_2	658 sec
Maneuver 3 (passenger ships only)		
Initial weight	$W_{0,3}$	408 t
Propellant supply (including 10% velocity reserve)	$W_{P,3}$	222 t
Final weight	$W_{1,3}$	186 t
Initial acceleration	$a_{0,3}$	0.49 g
Final acceleration	$a_{1,3}$	1.00 g
Burning time	t_3	298 sec
Maneuver 4 (passenger ships only)		
Initial weight	$W_{0,4}$	185 t
Propellant supply (including 10% velocity reserve)	$W_{P,4}$	134.5 t
Final weight	$W_{1,4}$	50.5 t
Initial acceleration	$a_{0,4}$	1.08 g
Final acceleration	$a_{1,4}$	2.84 g
Burning time	t_4	163 sec

21. DEPARTURE FROM THE ORBIT OF DEPARTURE (MANEUVER 1)

The length of the major semi-axis of both elliptical tracks to and from Mars is expressed (see Fig. 3) as

$$a = \tfrac{1}{2}(a_E + a_M) = 1.8875 \cdot 10^{13} \text{ cm.} \qquad (21.1)$$

The perihelian velocity of these ellipses level with earth's circum-solar orbit is found by equation (3.2) as

$$v_p = v_E \left(2 - \frac{a_E}{a}\right)^{\!\frac{1}{2}} = 32.83 \cdot 10^5 \text{ cm sec}^{-1}. \qquad (21.2)$$

The velocity of a Mars-bound ship, when leaving the field of earth's gravity, must therefore exceed the earth's velocity by

$$v_{d,1} = v_p - v_E = 3.03 \cdot 10^5 \text{ cm sec}^{-1}. \qquad (21.3)$$

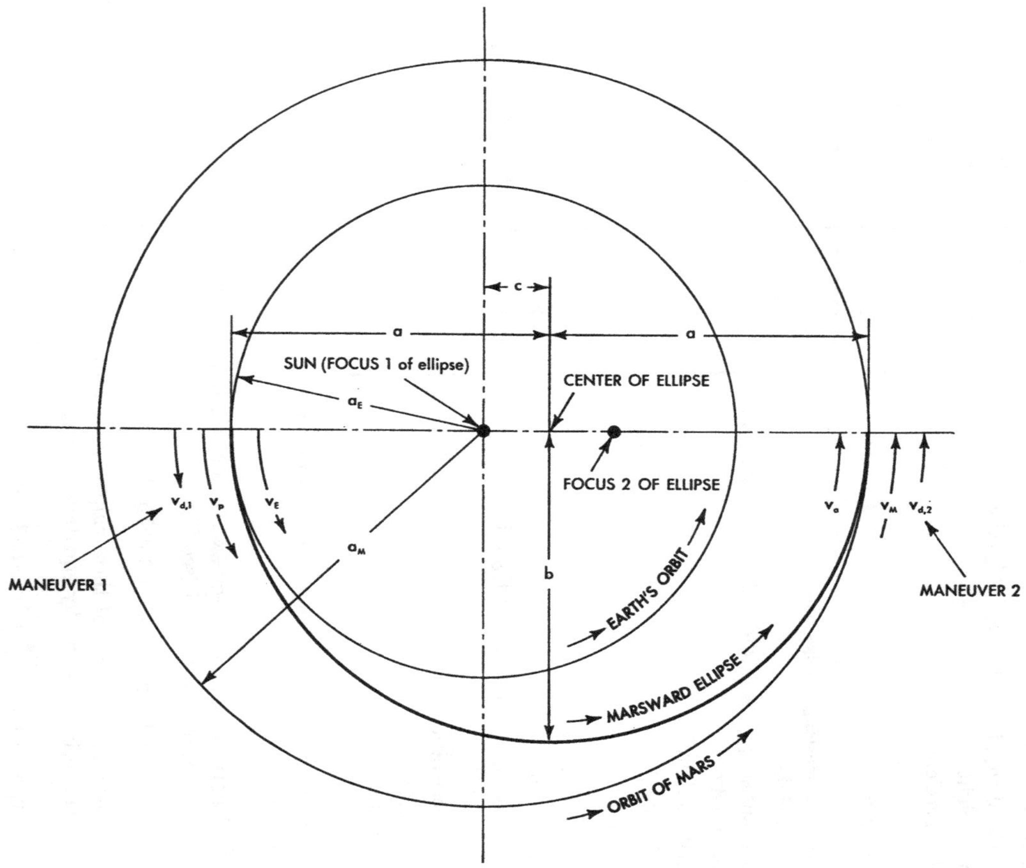

Figure 3. Marsward ellipse.

With this degree of residual velocity in the same direction as the earth's circum-solar motion, the ship will escape from the earth's field of gravity. The velocity increment required to bring about this result by a power maneuver in the orbit of departure may be determined as follows:

The kinetic energy per unit mass at departure from the circum-tellurian orbit of departure and as represented by the expression $\frac{1}{2} v_{tot,1}'^2$ must equal the sum of the residual energy $\frac{1}{2} v_{d,1}^2$ and the parabolic escape energy $\frac{1}{2} v_{p,1}^2$ referred level with the orbit of departure.

Since $v_{p,1}^2 = 2 v_{ci,1}^2$ is generally valid, we obtain

$$v_{tot,1}' = \sqrt{2 v_{ci,1}^2 + v_{d,1}^2} = 10.38 \cdot 10^5 \text{ cm sec}^{-1}. \quad (21.4)$$

On the other hand, the ship's velocity in the orbit of departure prior to maneuver 1 is already $v_{ci,1} = 7.07 \cdot 10^5$ cm sec^{-1}. Hence it is only necessary to bring up a velocity increment of $v_1' = v_{tot,1}' - v_{ci,1} = 3.31 \cdot 10^5$ cm sec^{-1} to reach the velocity of escape required to enter the desired circum-solar ellipse.

With the absolute velocity at thrust cut-off $v_{tot,1}'$ (as referred to the geocenter) the ship will immediately enter and coast through an escape hyperbola whose asymptote must lie parallel to the circum-solar orbital tangent of the earth if the ship is to follow the desired Mars-bound ellipse on leaving the field of earth's gravity. Hyperbolic tracks of this nature occur *after* departure thrust maneuvers 1 and 3, and also *before* arrival adaptation maneuvers 2 and 4.

a. Power track

In computing $v_{tot,1}'$ and v_1' an assumption was made, greatly simplifying the work, that the velocity increment v_1' took place instantaneously, i.e., that there was no work to be done against the potential of the tellurian field of gravity while the rocket motors were in operation. It may be seen that such a simplifying assumption is permissible when it applies to maneuvers 2, 3, and 4. In the latter maneuvers, accelerations have markedly increased and combustion periods and paths are relatively short in consequence. During maneuver 1 the weight of the ship is still very great compared to the relatively weak thrust of 200 tons. The simplified mode of computation will, in such a case, lead to considerable error, so the length of this power track must absolutely be considered.

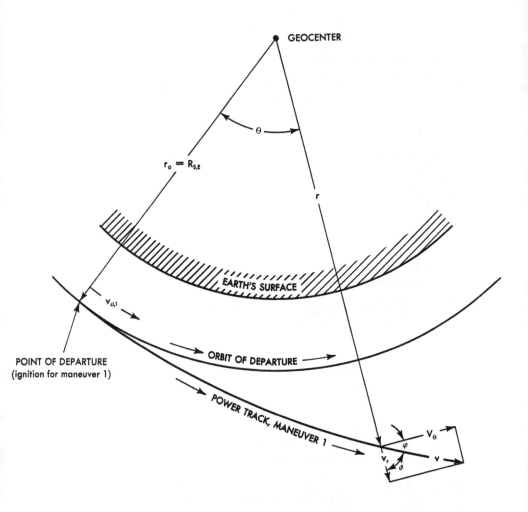

Figure 4. Computation of power track, maneuver 1.
See equations 21.5 to 21.18.

The power track for maneuver 1 was obtained by step-by-step integration as follows:

Lagrange's equation of energy states generally

$$\frac{\mathrm{d}}{\mathrm{d}t} \frac{\partial E}{\partial \dot{q}_i} - \frac{\partial E}{\partial q_i} = K_q \tag{21.5}$$

wherein q is a generalized coordinate, i is the subscript of the current coordinate number, and E is the energy.

For the kinetic energy we have (see Fig. 4):

$$E = \tfrac{1}{2} mv^2 = \tfrac{1}{2} \left[r^2\dot{\Theta}^2 + \dot{r}^2 \right]. \tag{21.6}$$

Hence

$$\frac{\partial E}{\partial r} = mr\dot{\Theta}^2; \quad \frac{\partial E}{\partial \dot{r}} = m\dot{r}; \quad \frac{\mathrm{d}}{\mathrm{d}t}\left(\frac{\partial E}{\partial \dot{r}} \right) = m\ddot{r} \tag{21.7}$$

$$\frac{\partial E}{\partial \Theta} = 0; \quad \frac{\partial E}{\partial \dot{\Theta}} = mr^2\dot{\Theta}; \quad \frac{\mathrm{d}}{\mathrm{d}t}\left(\frac{\partial E}{\partial \dot{\Theta}} \right) = mr^2\ddot{\Theta} + 2mr\dot{r}\dot{\Theta}. \tag{21.8}$$

For the forces we have

$$\delta \cdot A_r = F_r \cdot \delta_r \tag{21.9}$$

$$\delta \cdot A_\Theta = F_\Theta \cdot r \cdot \delta_\Theta \tag{21.10}$$

where $\delta \cdot A_r$, $\delta \cdot A_\Theta$ are the "virtual work" in the r and Θ directions, δ_r, δ_Θ the "virtual displacements," and F_r, F_Θ the generalized forces acting upon the ship in the direction indicated by a local plumb-bob and normal thereto.

Hence,

$$\ddot{r} - r\dot{\Theta}^2 = F_r/m \tag{21.11}$$

$$r\ddot{\Theta} + 2\dot{r}\dot{\Theta} = F_\Theta/m \tag{21.12}$$

where m represents the instananeous mass of the ship.

Let F be the rocket thrust. Then (see Fig. 4)

$$F_r = F \cos \vartheta - mg_r = F \cdot \frac{\dot{r}}{v} - mg_r \tag{21.13}$$

$$F_\Theta = F \sin \vartheta = F \cdot \frac{r\dot{\Theta}}{v} \tag{21.14}$$

where $g_r = g_0 R_E^2 / r^2$ is the local value of gravitational acceleration.

Thus we obtain the equations of motion for the calculation steps:

$$\ddot{r} = F \cdot \frac{\dot{r}}{mv} - g_r + r\dot{\Theta}^2 \tag{21.15}$$

$$r\ddot{\Theta} = F \cdot \frac{r\dot{\Theta}}{mv} - 2\dot{r}\dot{\Theta} \qquad (21.16)$$

$$v = \sqrt{(r\dot{\Theta})^2 + \dot{r}^2} \qquad (21.17)$$

It is herein assumed that the ship's gyroscopic autopilot maintains the direction of thrust, i.e., the vessel's longitudinal axis, tangent to the track. This is easily provided for by a program device.

Numerical integration must be initiated with conditions as imposed by the orbit of departure:

$$t_0 = 0; \quad r_0 = R_{S,E}; \quad \dot{r}_0 = 0; \quad \Theta_0 = 0; \quad \dot{\Theta}_0 = \frac{v_{ci,1}}{R_{S,E}}. \qquad (21.18)$$

The following data on the ship itself are used:

Initial weight: $W_{0,1} = 3.720 \cdot 10^9$ gm.
Propellant supply for maneuver 1: $W_{P,1} = 2.814 \cdot 10^9$ gm.
Propellant consumption per second: $\dot{W} = 0.702 \cdot 10^6$ gm.
Thrust: $F = 20 \cdot 10^7$ gm.

The resulting transients of the power track are compiled in Table 22, where $\Sigma\Delta t$ is the time elapsed from ignition, v the velocity, r the geocentric distance, Θ the polar angle measured against the direction indicated by a plumb-bob at the departure point, φ the flight path angle against the local horizontal.

TABLE 22. POWER TRACK MANEUVER 1

$\Sigma\Delta t$ (sec)	$v \cdot 10^{-5}$ (cm sec^{-1})	$r \cdot 10^{-8}$ (cm)	Θ (deg)	φ (deg)
0	7.07	8.11	0	0
500	7.32	8.13	25.6	1.4
1000	7.49	8.33	51.5	4.3
1500	7.53	8.72	76.7	8.5
2000	7.46	9.45	99.8	13.5
2500	7.36	10.49	120.2	19.0
3000	7.31	11.85	137.9	24.9
3500	7.42	13.56	152.6	31.0
4000	7.80	15.68	165.0	37.3

Note that there is a transient diminution of velocity v. This is due to the variable ratio of the vessel's acceleration to the retarding component of the acceleration of gravity.

The entire available period of combustion for maneuver 1 amounts to $t_1^* = W_{P,1}/\dot{W} = 4{,}008$ sec, and the terminal velocity obtainable by its application is found by continuing the above calculation. It is $v_{tot,1}^* = 7.805 \cdot 10^5$ cm sec^{-1}.

On the other hand, the required terminal velocity $v_{tot,1}$ has already been attained when the ship passes a distance of $r_1 = 15.56 \cdot 10^8$ cm from the geocenter. This occurs after a combustion period of $t_1 = 3{,}965$ sec. The polar angle at this instant is $\Theta_1 = 164.0°$ and the flight path angle $\varphi_1 = 36.8°$.

The considerable increase of potential energy during the thrust period renders $v_{tot,1}$ considerably smaller than the terminal velocity required by a short thrust maneuver in the orbit of departure. For the latter maneuver equation (21.4) provides $v'_{tot,1} = 10.38 \cdot 10^5$ cm sec^{-1}. Actually at the distance of $r_1 = 15.56 \cdot 10^8$ cm from the geocenter, appropriate use of (21.4) shows the terminal velocity required to be only $v_{tot,1} = 7.780 \cdot 10^5$ cm sec^{-1}.

The available propellant reserve of the first maneuver is equivalent therefore to a thrust reserve of the length of $t_{res,1} = t_1^* - t_1 = 43$ sec, and the corresponding velocity reserve is $v_{res,1} = v_{tot,1}^* - v_{tot,1} = 25 \cdot 10^2$ cm sec^{-1} or 3.5 per cent of the velocity change.

The power track and the ensuing escape hyperbola of maneuver 1 are shown in Fig. 5.

Remarks: It is demonstrable that maneuver 1 might be carried out at a somewhat lower initial weight of the vessel if one would increase the thrust somewhat. For example, four supplementary power plants of equal performance might be utilized. They would produce five times the thrust and be jettisoned after maneuver 1. Of course this expedient would increase the dead weight not only by the weight of the supplementary power plants themselves, but likewise because the other parts of the vessel would necessarily require reinforcement to withstand the higher accelerations. Nonetheless, the expedient would permit the use of so much less propellants that the initial weight of the vessels as ready to leave the orbit of departure would be reduced by about 10 per cent. The total transportation task imposed upon the ferry vessels would be eased by the same amount. This possible refinement has not been considered in the ensuing study.

b. Escape hyperbola

The ship enters the escape hyperbola through which it is to leave the earth's gravitational field at thrust cut-off of

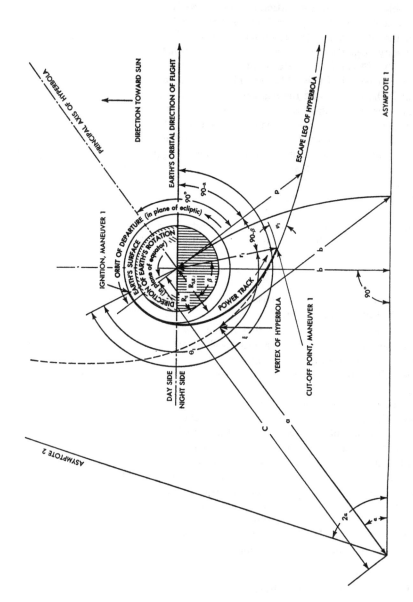

Figure 5. Maneuver of departure and escape hyperbola (maneuver 1).

maneuver 1. The data of this hyperbola are obtainable from the four characterizing values r_1, $v_{tot,1}$, Θ_1, and φ_1 just computed for the point of cut-off.

The polar equation of the hyperbola reads (see Fig. 5):

$$r_1 = \frac{p}{1 + e \cos \beta} \qquad (21.19)$$

wherein p is the parameter of the hyperbola, i.e., its focal ordinate, and β the required angle between the radius vector of the cut-off point and the principal axis of the hyperbola.

If the acceleration of gravity at the point of cut-off is written

$$g_1 = g_0 \frac{R_E^2}{r_1^2} = 165 \text{ cm sec}^{-2} \qquad (21.20)$$

then the length of the parameter will be

$$p = \frac{v_{tot,1}^2 \cdot \cos^2 \varphi_1}{g_1} = 23.5 \cdot 10^8 \text{ cm.} \qquad (21.21)$$

The numerical eccentricity is

$$e = \sqrt{1 + (v_{tot,1}^2 - 2g_1 r_1) \frac{v_{tot,1}^2 \cdot \cos^2 \varphi_1}{g_1^2 \cdot r_1^2}} = 1.239 \qquad (21.22)$$

and accordingly the linear eccentricity will be

$$c = \frac{e}{e^2 - 1} p = 54.45 \cdot 10^8 \text{ cm} \qquad (21.23)$$

the length of the major semi-axis being

$$a = \frac{1}{e^2 - 1} p = 43.9 \cdot 10^8 \text{ cm} \qquad (21.24)$$

and that of the minor semi-axis being

$$b = \sqrt{a \cdot p} = 32.1 \cdot 10^8 \text{ cm.} \qquad (21.25)$$

The angle between the escape asymptote and the principal axis is then

$$\tan \alpha = \frac{b}{a} = 0.731; \quad \alpha = 36° \, 10' \qquad (21.26)$$

and finally (using equation (21.19)) the angle β is given by

$$\cos \beta = \frac{1}{e} \left(\frac{p}{r_1} - 1 \right) = 0.412; \quad \beta = 65° \, 40'. \qquad (21.27)$$

The point of ignition of maneuver 1 is determined by the requirement that the escape asymptote must be parallel to the orbital motion of the earth. Figure 5 shows the geometric relation

$$\xi = \Theta_1 + 90 - \beta + 90 - \alpha = 242° \, 10'. \qquad (21.28)$$

22. MANEUVERS 2, 3, AND 4

The length of the power tracks may be neglected in first approximation when considering the above maneuvers.

a. Adaptation to circum-Martian orbit (maneuver 2)

The aphelian velocities of the Marsward and earthward ellipses are found by use of equations (5.1) and (21.2), as well as by Fig. 3, to be

$$v_a = \frac{a_E}{a_M} v_p = 21.55 \cdot 10^5 \text{ cm sec}^{-1}. \qquad (22.1)$$

At the aphelion the vessel will be overhauled by Mars with a velocity differential of

$$v_{d,2} = v_M - v_a = 2.55 \cdot 10^5 \text{ cm sec}^{-1}. \qquad (22.2)$$

The ship's aphelion occurs some thousands of kilometers this side of Mars's circum-solar orbit, and proper timing will bring it at this velocity into Mars's field of gravity. The latter will alter the ship's track to a hyperbolic loop whose focus is the Martian geocenter. The hyperbola selected should have a vertex distance $R_{S,M} = 4.390 \cdot 10^8$ cm from the Martian geocenter so that the vessel will enter a satellitic orbit 1,000 kilometers from Mars's surface. The circular velocity at this altitude is

$$v_{ci,2} = v_{p,M} \sqrt{\frac{R_M}{2R_{S,M}}} = 3.140 \cdot 10^5 \text{ cm sec}^{-1}. \qquad (22.3)$$

$v_{p,M} = 5.04 \cdot 10^5$ cm sec^{-1} is here the parabolic escape velocity at Martian ground level and R_M the radius of the Martian spheroid.

Appropriate application of equation (21.4) results in the velocity at the vertex of the hyperbola

$$v_{tot,2} = \sqrt{2v_{ci,2}^2 + v_{d,2}^2} = 5.15 \cdot 10^5 \text{ cm sec}^{-1}. \qquad (22.4)$$

Thus a retarding thrust must be applied to convert the hyperbolic track to a circular one, and the velocity difference required is

$$v_2' = v_{tot,2} - v_{ci,2} = 2.01 \cdot 10^5 \text{ cm sec}^{-1}. \qquad (22.5)$$

If we have envisaged a velocity reserve of 10 per cent for maneuver 2, mass ratio and propellant supply for this maneuver must be laid out for a velocity alteration of

$$v_2 = 2.21 \cdot 10^5 \text{ cm sec}^{-1}. \qquad (22.6)$$

The data of the hyperbola of adaptation are determined by the initial velocity $v_{d,2}$ and the selected distance of the hyperbolic vertex from the Martian geocenter.

To be determined are (see Fig. 6):

1. The asymptotic distance b with which the ship must approach Mars to enter an approach hyperbola having a vertex distance of $R_{S,M}$. The aphelion of the Marsward ellipse must occur *inside* the circum-solar Martian orbit by this distance b, so that the hyperbolic loop will be in the direction of Mars's rotation. This direction of circum-Martian revolution is desirable when the landing boats take off from Mars.

2. The angle α between the principal axis of the hyperbola and the approach asymptote.

The angle α is found by Pirquet's relation

$$\tan \alpha = \frac{b}{a} = 2 \cdot \frac{v_{d,2}^2}{v_{p,b}^2}. \qquad (22.7)$$

If we now tentatively select an asymptotic distance $b = 8.80 \cdot 10^8$ cm we find that the parabolic escape velocity at distance b is

$$v_{p,b} = v_{p,M} \sqrt{\frac{R_M}{b}} = 3.122 \cdot 10^5 \text{ cm sec}^{-1} \qquad (22.8)$$

and, with the use of equation (22.7), $\tan \alpha = 1.332$ and $\alpha = 53° 6'$. The asymptote of escape along which the vessel would escape from Mars's gravity if no retardation thrust were applied lies therefore at an angle of $2\alpha = 106° 12'$ to the asymptote of approach. The length of the major semi-axis is found to be $a = 6.60 \cdot 10^8$ cm with the help of equation (22.7).

The linear eccentricity is

$$c = \sqrt{a^2 + b^2} = 11.00 \cdot 10^8 \text{ cm}. \qquad (22.9)$$

The distance of the vertex from Mars's geocenter is then $R_{S,M} = c - a = 4.40 \cdot 10^8$ cm and corresponds with good accuracy to the desired value, indicating that the tentative value assumed for b was correct.

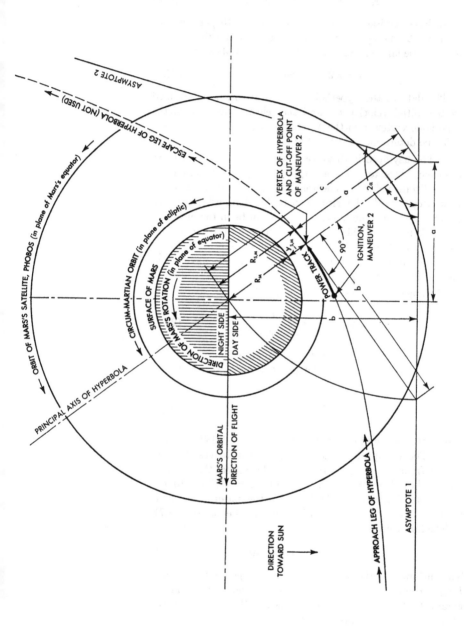

Figure 6. Adaptation to circum-Martian orbit (maneuver 2).

b. Departure from the circum-Martian orbit (maneuver 3)

Equations (22.1) to (22.6) are based on pure energy considerations and are hence equally valid for the return voyage from the circum-Martian orbit. Thus the entire velocity requirement for maneuver 3, including a velocity reserve of 10 per cent, will be

$$v_3 = v_2 = 2.21 \cdot 10^5 \text{ cm sec}^{-1}. \qquad (22.10)$$

The escape hyperbola ensuing from maneuver 3 is largely similar to the hyperbola of adaptation preceding maneuver 2 (see Fig. 6). The escape asymptote must lie parallel to Mars's orbital motion, but in the opposite direction so that the vessel is retarded with respect to Mars by the velocity decrement of $v_{d,3} = v_{d,2} = 2.55 \cdot 10^5$ cm sec^{-1}.

c. Adaptation to earth (maneuver 4)

Neglecting the finite length of the power track, it was found that in the orbit of departure the velocity change of $v_1' = 3.31 \cdot 10^5$ cm sec^{-1} would be required (equation (21.4)) so that the ship might have a residual velocity of $v_{d,1} = 3.03 \cdot 10^5$ cm sec^{-1} when leaving the earth's gravitational field. At the perihelion of the home voyage, the ship will now overhaul the earth from "astern" and enter tellurian gravity, having the same velocity difference $v_{d,4} = v_{d,1}$.

If we again envisage a 10 per cent velocity reserve, the mass ratio and propellant supply for this maneuver must be laid out for a velocity requirement of $v_4 = v_1' + 10\% = 3.64 \cdot 10^5$ cm sec^{-1}.

We still have to determine the asymptote distance b from the tellurian geocenter by which the perihelion of the homeward ellipse must lie outside earth's orbit. This is for the purpose of locating the vertex of the hyperbola of adaptation at a distance of $R_{S,E} = 8.11 \cdot 10^8$ cm from the geocenter. If we tentatively select $b = 27.75 \cdot 10^8$ cm, equation (22.8) yields for the parabolic escape velocity at distance b:

$$v_{p,b} = v_{p,E} \sqrt{\frac{R_E}{b}} = 5.36 \cdot 10^5 \text{ cm sec}^{-1}. \qquad (22.11)$$

Equation (22.7) then gives

$$\tan \alpha = 2 \cdot \frac{v_{d,4}^2}{v_{p,b}^2} = 0.640; \qquad \alpha = 32° 37'. \qquad (22.12)$$

The length of the major semi-axis is then

$$a = \frac{b}{\tan \alpha} = 43.35 \cdot 10^8 \text{ cm.} \qquad (22.13)$$

The linear eccentricity is

$$c = \sqrt{a^2 + b^2} = 51.45 \cdot 10^8 \text{ cm.} \qquad (22.14)$$

The difference $R_{S,E} = c - a = 8.10 \cdot 10^8$ cm agrees satisfactorily with the desired vertex distance. This is to say that the selected value for b was correct. There is a reason why the perihelion should lie at this distance *outside* the tellurian orbit, namely because the direction of flight through the hyperbolic vertex must be in the direction of the orbit of departure or identical with the earth's own rotation (see Fig. 7).

23. MASS RATIOS AND PROPELLANT SUPPLY FOR THE FOUR MANEUVERS

To determine the mass ratios, propellant quantities, and over-all weights of a ship for each individual maneuver, it is expedient to begin with the last one. As will later be substantiated, the weight of each of the seven ships which return to the circum-tellurian orbit will be $W_{1,4} = 50.5$ (metric) tons. This weight will prevail after the completion of maneuver 4, i.e., the propellants will be completely exhausted, but the weight includes all personnel and other cargo.

Applying the basic rocket equation

$$v = u \cdot \log_e \frac{W_0}{W_1} \qquad (23.1)$$

and the velocity requirement for maneuver 4 as computed in the previous section, namely $v_4 = 3.64 \cdot 10^5$ cm sec^{-1}, we find the initial weight at time of thrust application for the maneuver to be $W_{0,4} = 185 \cdot 10^6$ gm. Taking the difference between this and the ultimate weight we obtain $W_{P,4} = W_{0,4} - W_{1,4} = 134.5 \cdot 10^6$ gm for the amount of propellants which will be needed for maneuver 4.

The terminal weight after maneuver 3 exceeds the initial weight prior to maneuver 4 by the amount equivalent to the empty tanks jettisoned after maneuver 3. Similar weight differences also are found between the earlier thrust maneuvers, and it should be noted that the tanks used for earlier thrust maneuvers weigh much more than the later ones. The weight data compiled in Table 21 for maneuvers 4, 3, and 2 were computed in this manner. Thrust durations were based on the actual velocity changes as specified above, i.e., reserve

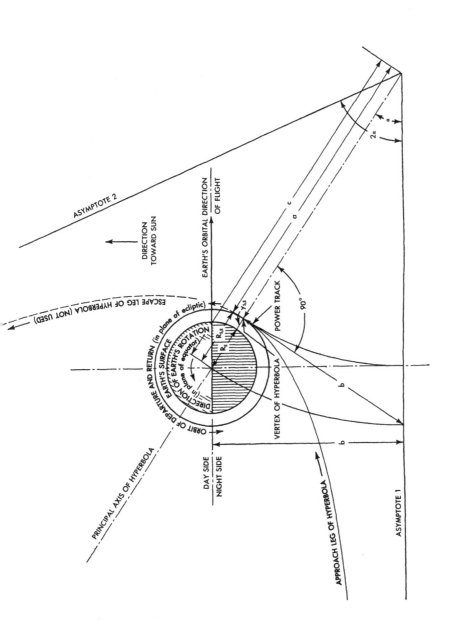

Figure 7. Adaptation to circum-tellurian orbit (maneuver 4).

propellants were not expended during main maneuvers. On the other hand, it has been assumed that such velocity reserves will be drawn upon between main thrust maneuvers in the form of corrective thrust periods, and thus each main maneuver was based upon a "theoretically correct" initial weight.

Weight data for maneuver 1 compiled in Table 21 are excerpted from the separate study found in section 21a.

Table 21 reveals that the terminal weight after maneuver 2 is still 410 tons when the vessel has entered the circum-Martian orbit, and this applies to both "passenger" and "cargo" vessels. The "cargo" ships have fulfilled their mission as soon as the landing boats have left them, and their remaining structures weigh but 15 tons. It is assumed that their habitable spaces during the interplanetary coast are those of the landing boats attached. Since the landing boats, as will later be shown, weigh but 200 tons apiece, it becomes obvious that a cargo vessel can have available out of the 410 tons arrival weight, 410 minus 15 minus 200 equals 195 tons for extra loads. Such extra cargo will include food, water, and oxygen rations for the entire expedition until time of departure from the circum-Martian orbit; furthermore, such items as spares, heavy radio equipment, a telescope for a thorough survey of the Martian surface before the descent of the landing boats; sounding missiles for a study of the Martian atmosphere from the circum-Martian orbit; "space-boats" for inter-ship visiting; and others.

24. VOLUMETRIC CONTENT OF TANKS AND THEIR DIMENSIONS

The volumetric requirements of the tanks constitute the main factors affecting the dimensions of the interplanetary vessels. They may be computed by referring to the specific weights of hydrazine and nitric acid as shown in section 14, and the required amounts of propellants for each of the four maneuvers as taken from Table 21. Figure 8 shows a line sketch of a possible disposition of the tanks of a "passenger vessel." The tanks containing the propellants for maneuvers 1 and 2 are shown located exteriorly to provide for easy detachment and jettisoning. If either of these maneuvers be completed without full exhaustion of the respective tanks, residual propellants may be transferred to the tanks marked "R" and arranged just outside the axial tanks for maneuver 4.

Table 23 shows diameter and length of the various tanks.

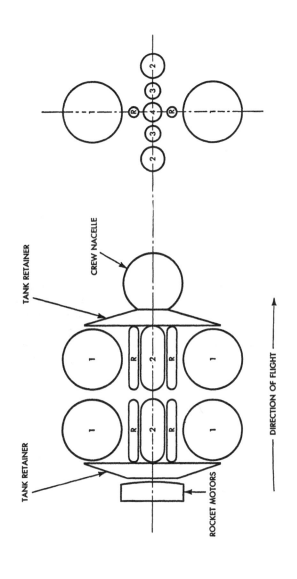

Figure 8. Tankage distribution of "passenger vessels." Figures denote the number of the corresponding maneuver. R = reserve tank.

TABLE 23. TANKS (PASSENGER VESSELS)

Maneuver	Number of hydrazine tanks	Number of nitric acid tanks	Diameter (m)	Length (m)
1	2	2	10.1	10.1
2	2	2	3.68	10.1
3	2	2	2.40	10.1
4	1	1	2.66	10.1
reserve	2	2	1.40	10.1

It is contemplated that the habitable space ("crew nacelle") of the passenger ships will be spherical, and 10 meters in diameter. With this, the configuration of the passenger vessels outlined in Fig. 8 will result in an over-all length of 41 meters and a maximum diameter of 29 meters. The cargo vessels, which have tankage for maneuvers 1 and 2 only, will bear in the location of the spherical habitable space a fully assembled, winged landing boat. This will increase the over-all length to about 64 meters on departure from the circum-tellurian orbit. Instead of the inner tanks for maneuvers 3 and 4, the cargo vessels will be equipped with large, silo-like containers for their 195 tons of extra payload. The maximum diameter of a cargo vessel will be determined by the wing span of the landing boat carried, and will depend on whether, during the interplanetary voyage, the wings are carried along fully assembled.

25. EMPTY WEIGHT OF PROPELLANT TANKS

Weights of propellant tanks for ships assembled and operated in space may be very low because of the small accelerations to which, particularly, the large tanks of the first maneuvers are subjected. An appropriate computation of the spherical tanks for maneuver 1 will exemplify this. According to Table 21, the initial acceleration in maneuver 1 is $a_{0,1} = .054\,g$. This would produce a pressure of $p_B = 1,010 \cdot \sigma_{NA} \cdot a_{0,1} = 86$ gm cm^{-2} at the bottom of the nitric acid tank.

Considering the container diameter envisaged, this pressure can be safely withstood by fabric-reinforced plastic of a weight of .065 gm cm^{-2}. The calculation shows for each tank a surface of $320 \cdot 10^4$ cm^2, making the weight of each of the four tanks for maneuver 1 equal to $20.8 \cdot 10^4$ gm. Each tank will require a ballonet or diaphragm to permit a light priming pressure to be exerted on its contents so that they may flow into the feed pumps just before the initiation of any thrust maneuver. After acceleration has begun, rearward pressure of propellants will exist until cut-off, but pump priming is otherwise impossible under weightless conditions. If this priming pressure is assumed to be 200 gm cm^{-2}, the weight of the tank material will be increased to .22 gm cm^{-2} and the weight of each tank to $70 \cdot 10^4$ gm. The four tanks for maneuver 1 will then weigh $2.8 \cdot 10^6$ gm.

As shown in Table 21, performance calculations were made under the assumption that there would be a diminution of weight amounting to $4 \cdot 10^6$ gm as a result of jettisoning the tanks for maneuver 1. The difference of $1.2 \cdot 10^6$ gm not only suffices for the retainers supporting the tanks but also for the helium banks for pressurizing them, as is shown by an over-all stress computation. Not only the tanks themselves, but also the helium banks and retainers are subject to being jettisoned.

No weight allowance has been made in this study for any kind of metallic armor shell ("meteor bumper") to protect the plastic containers from being punctured by small meteors. It has been assumed, instead, that the propellant tanks are self-sealing like gasoline tanks for military aircraft. Use of meteor bumpers of a wall thickness adequate to give a reasonable degree of protection against puncture during the lifetime of each tank would increase the dead weight of the ships only slightly. If such bumpers were added to the tanks of the ships described herein, the loss in performance would be in the order of one tenth of the velocity reserves provided for each maneuver.

26. ELAPSED TIMES ON THE MARSWARD AND EARTHWARD ELLIPSES

Equation (21.1) gave the length $a = 1.8875 \cdot 10^{13}$ cm of the major semi-axes of Marsward and earthward ellipses. The linear eccentricity is then (see Fig. 3)

$$c = a - a_E = 0.3925 \cdot 10^{13} \text{ cm.} \qquad (26.1)$$

The numerical eccentricity is

$$e = \frac{c}{a} = 0.208. \qquad (26.2)$$

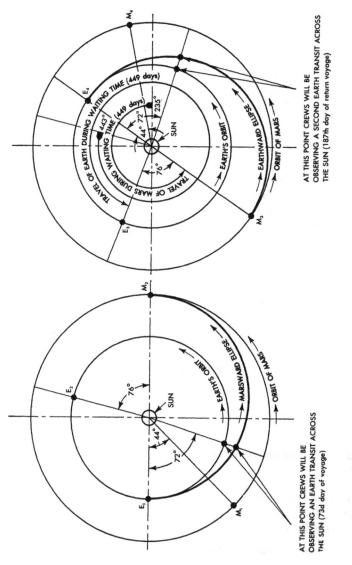

Figure 9. Scheduling of Mars expedition.

a. Marsbound voyage

E_1, M_1: Position of earth and Mars at time of departure from earth (maneuver 1).

E_2, M_2: Position of earth and Mars at arrival in circum-Martian orbit (maneuver 2).

b. Earthbound voyage

E_3, M_3: Position of earth and Mars at time of departure from circum-Martian orbit (maneuver 3).

E_4, M_4: Position of earth and Mars at arrival on earth (maneuver 4).

The length of the minor semi-axis is

$$b = \sqrt{a^2 - c^2} = 1.848 \cdot 10^{13} \text{ cm.} \qquad (26.3)$$

According to Kepler's third law and the data on earth's circum-solar orbit, the period of revolution through the entire ellipse will be

$$T = T_E \sqrt{\frac{a^3}{a_E^3}} = 44.8 \cdot 10^6 \text{ sec} = 520 \text{ days.} \qquad (26.4)$$

Half this, or the time of an earthward or Marsward trip, will be

$$t = 260 \text{ days.} \qquad (26.5)$$

27. SCHEDULING

The instant of departure from the circum-tellurian orbit is determined not only by conditions outlined in equation (21.28) but also by the fact that the space ships must arrive in the Martian circum-solar orbit when Mars is at that point thereof which contacts the aphelion of the Marsward ellipse. To simplify consideration of this problem, we shall again assume that the circum-solar orbits of both earth and Mars are truly circular, although actually the Martian orbit is markedly eccentric (see Table 20). It is also herein assumed that the angular velocities of the two planets are constant. Mars completes a circum-solar voyage of 360° every 687 days, covering a sector of .524° per day. Earth completes her circle around the sun in 365 days at a rate of .987° per day. The space ships will coast through the Marsward ellipse for 260 days; hence at the moment they leave the orbit of departure Mars must be .524 · 260 = 136° behind and be approaching the intersection of his orbit by the Marsward ellipse. As seen from the sun, this point of intersection will be diametrically opposed to the origin of the ellipse in earth's circum-solar orbit and therefore Mars, at the instant of the ships' departure, must be 180° − 136° = 44° ahead of earth (see Fig. 9a).

During the voyage along the Marsward ellipse, earth will cover an angular distance of .987 · 260 = 256°, and when the ships reach Mars, earth will be 256° − 180° = 76° ahead of the latter. It will be noted that earth will angularly overhaul the space ships on the 73d day of their voyage Marsward when they have coasted through a radius vector angle of 72°, and pass between them and the sun. Since the voyaging ellipse, Mars's orbit, and that of the earth lie with good accuracy in the plane of the ecliptic, the interplanetary crews will have the opportunity of observing earth transit across the face of the sun on the 73d day (see Fig. 9a).

The earthward trip will also require 260 days and so earth must, at the time of departure from Mars, stand 256° before the point where the perihelion of the earthward ellipse contacts earth's circum-solar orbit, that is to say that earth must be 256° − 180° = 76° behind Mars. On the homeward voyage, the crews will observe an earth transit across the sun on the 187th day of the trip (see Fig. 9b).

It is apparent that a particular relationship must exist between the positions of Mars and the earth for a successful earthward departure from the circum-Martian orbit, and that the interplanetary vessels circling there are obliged to "wait" for a definite period. This period is determined by the time required for earth to move from a position 76° ahead of Mars to one 76° behind him. Earth moves angularly faster around the sun and must traverse an angular distance of 360° − 2 · 76° = 208° with respect to Mars. The difference in angular velocities of the two planets is .987° − .524° = .463° per day. Thus the "waiting period" is 208/.463 = 449 days. During this time Mars traverses an arc of 449 · .524° = 235°, while earth traverses 449 · .987° = 443° (see Fig. 9b).

28. USEFUL LOADS AND EMPTY WEIGHTS

Section 23 provides proof that the three cargo space vessels are capable of transporting not only three fully loaded landing boats, but likewise a supplementary load of about 195 tons apiece. During the time before the departure of the seven passenger vessels on the earthward trip, the entire expedition can utilize this supplementary useful load of the three cargo vessels. Immediately prior to earthward departure, however, each of the seven passenger vessels will be required to embark and subsist 10 men apiece. This will require food, oxygen and water, and personal baggage for the 260 days of the return voyage, besides imposing upon them additional loads which may consist of Martian objects collected by the landing party. The required useful loads and an estimate of the weights of the principal structural components of a passenger vessel (after the tanks for maneuver 3 have been jettisoned) are shown in Table 24.

TABLE 24. BILL OF WEIGHTS OF PASSENGER VESSELS AFTER MANEUVER 3

	Metric tons
Ship	
Structure	1.0
Rocket motors	4.7
Turbopumps	1.0
Hydrogen peroxide	2.0
Steam plant, empty	0.2
Tubing and valves	0.2
Actuators for hinged rocket motors	0.6
Flywheels for spatial attitude control	1.0
Electric power supply	2.0
Crew nacelle, empty (plastic)	2.5
Nacelle equipment	1.5
Air lock and landing berth for space boats	0.8
Air recuperation plant	1.5
Water recuperation plant	1.5
Guidance system	1.5
Navigation equipment	1.0
Instrumentation and annunciators	0.5
Radio equipment for moderate range	1.0
Tanks for maneuver 4	0.5
Total	25.0
Payload	
10 crew members	0.8
Personal baggage for 260 days	1.2
Oxygen (1.235 kilograms per man per day, plus reserves)	5.0
Food (1.2 kilograms per man per day, plus reserves)	3.5
Food wrappings	1.0
Potable water (2 kilograms per man per day, plus reserves)	5.5
Utility water, initial weight*	2.0
Space boat including propellants	3.5
Space suits	0.5
Spares and tools	1.0
Unassigned payload	1.5
Total	25.5
Ship's weight after maneuver 3, but excluding propellant supply for maneuver 4	50.5

* Non-potable utility water recovered from the crew nacelle increases as the potable water decreases, the recovery amounting to some 1.6 kg per person per day. It is a result of exhalation and perspiration and may be re-used after passing through the desiccators and sterilizers.

There will be an extra load requirement in the matter of hydrogen peroxide during the lengthy thrust periods of the first maneuvers of the passenger vessels. Since the cargo vessels are still available between these maneuvers, this can largely be taken care of by transferring to the cargo vessels suitable quantities of payload listed in Table 24.

C

LANDING BOATS

29. STATEMENT OF PROBLEM

The landing boats must be capable of descending to Mars from the circum-Martian orbit and of re-ascending without replenishing their propellants. Wings may be used for the descent, considering the existence of a Martian atmosphere. There are, however, two complicating factors which differentiate this procedure from the not dissimilar technique of landing the third stage of the ferry vessels on earth. The factors are:

1. The density of the Martian atmosphere at ground level is only about 1/12 of the tellurian atmosphere, and the lift of any given wing is diminished in the same proportion. This, however, is partially compensated by the low acceleration of Martian gravity which is only .38 g_0. Thus the boat weighs less on Mars and requires far less lift.

2. The third stages of the ferry vessels land with empty tanks, while two Martian landing boats must carry enough propellants to re-ascend into the orbit. It will be shown below that such an ascent of a single-stage rocket is made relatively easy by the feeble Martian field of gravity.

The above two factors require, nevertheless, that extremely large wing areas be used. They and the landing gear will be stripped off before the ascent to the orbit. Thus only the torpedo-like hulls will return to the passenger ships which await them in the circum-Martian orbit.

In addition to the "landing party," the landing boats must carry means for subsisting the men for more than a year on Mars and for carrying out their scientific investigations. A majority of the personnel of the expedition will make the descent, while a certain number will remain aboard the circling passenger vessels as "shipkeepers." Most of the equipment can, of course, be abandoned on the surface of Mars, as can such consumables as have not been utilized. It will be shown that the total useful load with which the boats will land on Mars is 149 metric tons. It will suffice for 50 persons to remain on Mars for over 400 days. It includes rations, vehicles, inflatable rubber housing, combustibles, motor fuels, research equipment, tools, and the like.

Considering the risk attending a wheel landing on completely strange territory at relatively high speed, it is assumed that the first landing boat will make contact with the Martian surface on a snow-covered polar area, and on skis or runners, minimizing this risk. This boat will be abandoned on Mars because of the impossibility of re-ascending from polar latitudes to an orbit in the plane of the ecliptic. So this boat need not carry any fuel for the re-ascent and thus almost its entire useful load may be devoted to cargo. The total useful load of the ski boat will be shown to be 125 tons, part of which may well consist of ground vehicles. With such vehicles, the crew of the first landing boat would proceed to the Martian equator and there select or prepare a suitable strip for the wheeled landing gears of the remaining two boats. At the termination of work on Mars, the crew of the first boat would return in the remaining two boats to the waiting passenger ships.

It is desirable to avoid the difficult problem of assembling landing boats in the circum-tellurian orbit of departure. This may be done by mounting the boats atop the second stages of regular ferry vessels, replacing the usual third stage, and being given the launching impulse by the usual first stage of a ferry ship. The boats would be without their wings, which would be attached in the orbit of departure, having been delivered there disassembled. The boat with skis or runners would be equipped with supplementary propellant tanks of fabric for its ascent to the circum-tellurian orbit of departure. These extra tanks would be removed before Marsward departure, leaving interior space for stowage of the additional equipment carried by this boat.

30. SYMBOL MEANINGS AND TABULATION OF MAIN DATA

TABLE 25. MANEUVER OF DESCENT IN CIRCUM-MARTIAN ORBIT

Thrust	F	200 t
Initial weight	W_0	200 t
Final weight (landing weight, referred to 1 g_0)	W_1	185 t
Rate of propellant flow	$\dot{W}$	755 kg/sec
Exhaust velocity	u	2,600 m/sec
Landing payload:		
Ski boat	$W_{N,S}$	125 t
Wheeled boat	$W_{N,W}$	12 t
Burning time	t_{01}	17.2 sec
Velocity decrement for maneuver of descent	v_{01}	173 m/sec
Altitude of perigee of landing ellipse	y_p	155 km
Landing speed	v_L	196 km/hr

TABLE 26. MANEUVER OF RE-ASCENT TO CIRCUM-MARTIAN ORBIT (WHEELED BOATS ONLY)

Thrust	F	200 t
Initial weight (referred to 1 g_0)	W_2	138 t
Empty weight without payload (stripped of wings and landing gear)	W_E	20 t
Final weight	W_3	26.8 t
Payload of ascent	$W_{N,A}$	5 t
Ratio of empty weight to total propellant load	k	0.156
Propellant supply for maneuver of re-ascent	$W_{P,23}$	111.2 t
Take-off acceleration (relative)	a_{to}	10.5 m/sec^2
Burning time	t_{23}	147 sec
Cut-off altitude	y_3	125 km
Cut-off velocity	v_3	3,700 m/sec

TABLE 27. MANEUVER OF ADAPTATION (WHEELED BOATS ONLY)

Thrust	F	200 t
Initial weight	W_3	26.8 t
Final weight	W_4	25 t
Burning time	t_{34}	2.4 sec
Velocity increment for adaptation	v_{34}	180 m/sec

TABLE 28. DIMENSIONS

Length of body	L	22 m
Body diameter	D	4.7 m
Wing area	S_W	2,810 m²
Wing span	b_W	153 m
Total propellant supply (wheeled boats)	$W_{P,W}$	128 t

31. THE NATURE OF THE MARTIAN ATMOSPHERE

The mass of a column of Martian atmosphere standing above a square centimeter of surface is equal to 22 per cent of a column of tellurian air of identical section, according to measurements of Mars's reflectivity ("albedo"). Under the simplifying assumption that the compositions of Martian and tellurian air are similar and that temperatures are equal, the density of Martian air at ground level will be

$$\gamma_{0,M} = 0.22 \cdot \frac{g_M}{g_0} \cdot \gamma_{0,E} = 1.1 \cdot 10^{-4} \text{ gm cm}^{-3} \qquad (31.1)$$

and the pressure at ground level

$$p_{0,M} = 0.22 \cdot \frac{g_M}{g_0} \cdot p_{0,E} = 64 \text{ mm Hg} = 84 \text{ gm cm}^{-2}. \qquad (31.2)$$

At this pressure the boiling point of water is 44° C. The following law is derived from equation (11.2) and applies to the stratification of pressure and density of the Martian atmosphere:

$$\frac{\gamma}{\gamma_{0,M}} = \frac{p}{p_{0,M}} = e^{-\frac{0.38y}{y^*}} = 10^{-\frac{y}{47.15 \cdot 10^5 \text{ cm}}}. \qquad (31.3)$$

It is necessary to ascend 18 km in earth's atmosphere to encounter a reduction of atmospheric pressure by a factor of 10, while the same reduction in the Martian atmosphere will be found only by 47 km altitude increase. This is caused by the feeble Martian gravity which is unable to compress this atmosphere sufficiently. This results in the almost paradoxical conclusion that Mars's atmosphere above altitudes of 30 km is denser than that of earth at the same altitude, although the pressure at ground level is only 1/12 that of earth's. The condition is significant for the selection of the perigeal altitude of the landing maneuver.

32. WING AREAS AND LANDING SPEEDS

Let the landing weight of a boat, referred to an acceleration of gravity of $1\ g_0$, be $W_1 = 185 \cdot 10^6$ gm. The actual weight on Mars will, however, be only $.38 \cdot 185 \cdot 10^6 = 70.3 \cdot 10^6$ gm. If the design landing speed is then $v_L = 196$ km/hr $= 54.5 \cdot 10^2$ cm sec^{-1}, the required wing area according to equations (7.1) and (31.1) will be

$$S_W = \frac{2 \cdot g_M \cdot W_1}{c_{L,max} \cdot \gamma_{0,M} \cdot v_L^2} = 2.81 \cdot 10^7 \text{ cm}^2. \qquad (32.1)$$

33. SELECTION OF THE ALTITUDE OF THE PERIGEE OF THE LANDING ELLIPSE

The altitude of the perigee of the landing ellipse must be selected so that there is already adequate atmospheric density to exert a negative lift on the wings of the landing boat which will force the latter from its landing ellipse and into a circular path.

An investigation of this subject is similar to that for the third stage of a ferry vessel (see section 8). Applying equations (31.1), (31.3), and (32.1), a suitable perigeal altitude will be $y_p = 0.155 \cdot 10^8$ cm. This will be equivalent to a distance from Mars's geocenter of $R_p = 3.545 \cdot 10^8$ cm. The perigeal velocity at which the boat will pass into its glide through the Martian atmosphere is $v_p = 3.67 \cdot 10^5$ cm sec^{-1}.

34. INITIAL WEIGHT AT DEPARTURE FROM CIRCUM-MARTIAN ORBIT

The landing weight of a boat has hitherto arbitrarily been taken as $W_1 = 185 \cdot 10^6$ gm. How heavy is it, then, when it departs from the circum-Martian orbit? Using equation (5.1) the velocity at the apogee of the landing ellipse is $v_a = 2.967 \cdot 10^5$ cm sec^{-1}. Now the velocity in the circum-Martian orbit is $v_{ci,2} = 3.14 \cdot 10^5$ cm sec^{-1} so the boat must retard itself by a velocity decrement of $v_{01} = v_{ci,2} - v_a = 0.173 \cdot 10^5$ cm sec^{-1} to be able to enter the landing ellipse. According to the basic rocket equation (23.1), the initial weight will be $W_0' = 1.069 \cdot 185 \cdot 10^6 = 198 \cdot 10^6$ gm; in round figures $W_0 = 200 \cdot 10^6$ gm.

This computation was based upon an exhaust velocity of only $u = 2.6 \cdot 10^5$ cm sec^{-1}. The pure space vessels operating solely *in vacuo* can expand their discharge gases to a nozzle exit pressure of $p_e = 10$ gm cm^{-2}, while the nozzle exit pressure of the landing boats was set at $p_e = 66$ gm cm^{-2} in view of the requirement that they take off from the surface of Mars, and the risk of jet separation at the there existing pressure of $p_{0,M} = 84$ gm cm^{-2}. The design of a slimmer hull for the landing boats than for the third stage of a ferry vessel is also explained by this increased nozzle exit pressure. The propellant consumption per second of the landing boats is somewhat higher than that of a space ship proper or a ferry third stage, despite equal thrusts. It is inversely proportional to the lower exhaust velocity, amounting to $\dot{W} = 0.755 \cdot 10^6$ gm sec^{-1}. Propellants used for the landing maneuver are equivalent to $W_{P,01} = W'_0 - W_1 = 13 \cdot 10^6$ gm and the combustion period is $t_{01} = W_{P,01}/\dot{W} = 17.2$ sec.

35. REGAINING THE CIRCUM-MARTIAN ORBIT

A rough stress calculation reveals that the wings called for in equation (32.1) will weigh some $W_W = 33 \cdot 10^6$ gm referred to 1 g_0. The use of sheet steel for wing skinning is assumed, and an adequate safety factor for the aerodynamic forces actually occurring is included. If the landing gear weight is $W_G = 7 \cdot 10^6$ gm, it is apparent that $40 \cdot 10^6$ gm of the total landing weight of $W_1 = 185 \cdot 10^6$ gm can be stripped off and be abandoned on Mars. If the useful load for ascent to the circum-Martian orbit be further restricted to $W_{N,A} = 5 \cdot 10^6$ gm, as compared to the landing useful loads of the wheeled boats of $W_{N,W} = 12 \cdot 10^6$ gm, there is a further weight reduction of $7 \cdot 10^6$ gm. Thus the take-off weight from Mars of each of the two boats is, accordingly, $W_2 = 138 \cdot 10^6$ gm referred to 1 g_0. Applying a thrust of $F = 20 \cdot 10^7$ gm, we get a relative take-off acceleration

$$a_1 = \frac{F \cdot g_0}{W_2} - g_M = 10.5 \cdot 10^2 \text{ cm sec}^{-2}. \qquad (35.1)$$

Let it be assumed that at altitude $y_3 = 0.125 \cdot 10^8$ cm the boat will reach its maximum velocity traveling horizontally to Mars's surface and that it will enter the ellipse of ascent to the circum-Martian orbit with this velocity. A calculation involving equation (3.2) gives a perigeal velocity in the ellipse of ascent at this altitude of $v_3 = 3.70 \cdot 10^5$ cm sec^{-1}. The apogeal velocity of this ellipse is found by equation (5.1) to be $v_a = 2.96 \cdot 10^5$ cm sec^{-1}, requiring that a velocity increase of $v_{34} = v_{ci,2} - v_a = 0.18 \cdot 10^5$ cm sec^{-1} be achieved to perform a maneuver of adaptation in the circum-Martian orbit. With the help of the basic rocket equation (23.1) we find $W_3 = 1.072\,W_4$.

W_4 represents the weight of the empty, wingless boat plus its useful load for the ascent, $W_{N,A} = 5 \cdot 10^6$ gm. If $W_4 = 25 \cdot 10^6$ gm, the initial weight for the maneuver of adaptation becomes $W_3 = 26.8 \cdot 10^6$ gm and this is, coincidentally, the terminal weight for the maneuver of ascent. Hence the mass ratio available for the ascent is

$$E_{23} = \frac{W_2}{W_3} = 5.15. \tag{35.2}$$

By utilizing equation (4.2) it is easy to determine roughly whether the required perigeal velocity $v_3 = 3.70 \cdot 10^5$ cm sec^{-1} is actually attainable with this mass ratio. It is, of course, essential to substitute for the acceleration of gravity g used in (4.2) the corresponding figure $g_M = 0.38 \cdot 981 = 373$ cm sec^{-2} applicable on Mars. If a mean track angle of inclination of $\varphi_m = 45°$ is assumed for purposes of simplification, equation (4.2) solves for a velocity at power cut-off of $v_3^* = 3.877 \cdot 10^5$ cm sec^{-1}. Equation (4.4), having been appropriately modified to apply to such a 45° ascent in the Martian field of gravity, would show the altitude of cut-off to be $y_3^* = 0.145 \cdot 10^8$ cm, and for the combustion period for this ascent we find

$$t_{23} = \frac{W_2 - W_3}{\dot{W}} = 147 \text{ sec.}$$

This rough calculation indicates that there is no difficulty in attaining the required perigeal velocity at the proposed altitude with the wingless boat. It would appear that an exact numerical integration of the track of ascent in accordance with the procedure presented in section 4 should result even more favorably, for it considers not only aerodynamic drag but also such alleviating factors as decrease of gravity with increase of altitude, actual track angle variations, peripheral velocity at the Martian equator, centrifugal decrement, and gains in static thrust due to decrease of ambient atmospheric pressure. Reserves herein discovered might be applied to an increase of payload.

36. USE OF WINGLESS HULLS OF LANDING BOATS AS THIRD STAGES OF FERRY VESSELS

The weight of an empty landing boat, less wings but including useful loading $W_{N,A} = 5 \cdot 10^6$ gm as applicable to departure from the surface of Mars, is $W_4 = 25 \cdot 10^6$ gm. The wheeled boats, however, will descend to Mars bearing a payload heavier by $7 \cdot 10^6$ gm. Therefore the terminal weight of a fully loaded but wingless boat after arrival in the circum-tellurian orbit of departure will be $W^*_{2,\text{III}} = 32 \cdot 10^6$ gm.

The initial weight of a boat when mounted atop the first two stages of a ferry vessel must be equal to the initial weight of the normal third stage to conform with the performance of the regular three-stage ferry ship. The weight is then $W_{0,\text{III}} = 130 \cdot 10^6$ gm.

A comparison of the terminal weight $W^*_{2,\text{III}} = 32 \cdot 10^6$ gm with the terminal weight $W_{2,\text{III}} = 66.6 \cdot 10^6$ gm of a regular third stage of a ferry vessel (after the maneuver of adaptation) shows the landing boat to possess great reserves in the matter of mass ratio (see Table 5). Thus the landing boat ascending to the circum-tellurian orbit can even carry an extra propellant payload.

37. DIMENSIONS

If it is desired to hyperexpand the exhaust gases to 78 per cent of the ambient atmospheric pressure when taking off from Mars, the nozzle exit area of a landing boat will be computed as $176 \cdot 10^3$ cm² and the combined diameters of the exhaust nozzles as $D = 4.7 \cdot 10^2$ cm. Let the diameter of the boat hull be identical with this figure.

Table 28 shows that a wheeled boat must have tankage commensurate with a propellant weight of $W_{P,W} = 128 \cdot 10^6$ gm. Equation (14.1) applied hereto calls for a tankage volume of $99 \cdot 10^6$ cm³. If the inside diameter of a tank be 440 cm, the required length of the cylindrical tank space will be 650 cm. We may here apply the same estimates as were used in the case of the ferry vessels, namely: length of rocket motors, 300 cm; allowance for domed tank headers, 100 cm; length of pump and valve compartment, 200 cm. This makes the power section of the boat 1,250 cm long. Let 950 cm be assigned to such spaces as pilot's cabin, radio space, cargo space, and automatic guidance compartment. Total length of the hull is then $L = 22$ meters. If this hull be equipped with backswept, all-wing surfaces, their span will be about $b_S = 153$ meters.

38. GLIDE PATH FROM PERIGEE OF LANDING ELLIPSE TO GROUND CONTACT

The configuration and length of the glide path of a landing boat may be determined in accordance with the procedure given in section 11, including the use of equation (31.1) which defines the density stratification of the Martian atmosphere. This computation is based upon the aerodynamic assumptions used in section 10, a wing area corresponding to the requirements of equation (32.1), and a landing weight of the boat equivalent to $W_1 = 185 \cdot 10^6$ gm. The result is shown in Table 29.

TABLE 29. GLIDE PATH OF LANDING BOAT

Velocity $v \cdot 10^{-5}$ (cm sec^{-1})	Altitude $y \cdot 10^{-5}$ (cm)	Deceleration $-\dot{v} \cdot 10^{-2}$ (cm sec^{-2})	Time span Δt (sec)	Distance proceeded $\Delta x \cdot 10^{-5}$ (cm)	Rate of descent $\dot{y} \cdot 10^{-2}$ (cm sec^{-1})	Total time $\Sigma \Delta t$ (sec)	Total distance $\Sigma \Delta x \cdot 10^{-5}$ (cm)
3.67	155						
		0.183	930	3,330	0	930	3,330
3.5	155						
		0.336	1,490	4,830	14.8	2,420	8,160
3	133						
		0.955	1,046	2,620	32.4	3,466	10,780
2	99						
		1.105	905	1,360	18.7	4,371	12,140
1	82						
		0.867	808	525	53.3	5,179	12,665
sonic speed	39						

The elapsed time of flight from the perigee of the landing ellipse to the moment sonic speed is reached is, then, 5,179 sec = 1 hr 26 min 19 sec, and the glide track preceding attainment of sonic speed is 12,665 km long, or about 60 per cent of Mars's circumference. The glide is subsonic beginning at an altitude of 39 km, and this is extremely helpful in the search for a suitable landing spot. The preceding calculation also indicates that no difficulty will be encountered in reaching one of Mars's polar caps from a landing ellipse in the plane of the ecliptic.

39. SKIN TEMPERATURE INCREASE DURING THE GLIDE OF A LANDING BOAT

Section 12 included a study of the skin temperature of a ferry vessel during its glide to earth. The same mathematical procedure applied to the glide of a landing boat through the Martian atmosphere as computed in section 38 gives results which are shown in Table 30.

TABLE 30. SKIN HEATING DURING GLIDE OF LANDING BOAT

$v \cdot 10^{-5}$ (cm sec^{-1})	$y \cdot 10^{-5}$ (cm)	T_s (°K)
3.67	155	643
3.5	155	625
3	133	649
2	99	605
1	82	416

It is apparent that the skin temperatures remain within such low limits that the use of light alloys for skinning might almost be considered.

FERRY FLIGHTS AND GENERAL LOGISTICS

For any expedition to Mars to be successful, it is essential that the first phase of space travel, the development of a reliable ferry vessel which can carry personnel into a circum-tellurian orbit, be successfully completed. It was proved under Heading a that such a ferry vessel can be built, and that it can carry substantial payloads into the circum-tellurian orbit, that its first and second stages can be salvaged and re-used, and that its third stage can make a normal glider landing. It is not intended to develop herein the scope or the cost of the development of such ferry vessels. Rather is it assumed in the following that the ferry vessel has already attained a degree of reliability comparable to that of modern transport aircraft.

40. FERRY FLIGHTS

Each of the ten interplanetary vessels departing from the circum-tellurian orbit weighs initially $W_{0,1} = 3,720$ (metric) tons (see Table 21). The ferry vessels will be required, therefore, to transport cumulative payloads of 37,200 tons into that orbit. The payload of each ferry vessel is $W_{N,III,02} = 39.4$ tons (see Table 8). If we assume this payload to be but 39 tons on the average, then the preparation of the expedition to Mars will require 37,200/39 = 950 ferry flights.

Each ferry flight consumes 5,583 tons of propellants (see Table 8), so that the whole ferry operation will need about 5,320,000 tons. This is equivalent to the capacity of 443 tank ships of 12,000 tons displacement each. It is interesting to compare this to the official statistics which show that about 10 per cent of an equivalent quantity of high octane aviation gasoline was burned during the six months' operation of the Berlin Airlift.[1]

[1] See Aeronautical Engineering Review, March, 1949, p. 25 ff.

If we count on mass-produced hydrazine and nitric acid, in the required proportion, costing 100 dollars per ton on the average, the cost of the propellants for the ferry flights would be around 500 million dollars. The propellants for the actual interplanetary voyage are trifling in comparison, being no more than 3.5 million dollars.

Regarding the time element of the ferry operation, let us assume that each ferry vessel can carry out a round trip to the orbit every ten days. This interval seems attainable if we allow three days for salvage and return to base of the second stage after its landing in the ocean 1,459 km from the take-off point. This would leave seven days for inspection, reconditioning, and reassembly of the three stages. On that basis, 46 ferry vessels could accomplish the 950 flights in 8 months, even if 6 of the vessels were continuously out of commission. Three of the vessels only would be utilized for "dry cargo"; the remainder would carry propellants only.

41. LOGISTIC COMPARISONS

Table 31 gives the most important data on the whole expedition, and on its capabilities and logistics. They are presented in convenient form for comparison.

TABLE 31. TOTAL ENTERPRISE AND GENERAL LOGISTICS

Crew members	70
Total duration of expedition	2 years 239 days
Voyaging time, earth to Mars	260 days
Waiting time in circum-Martian orbit	449 days
Hereof, stay on surface of Mars (approximately 50 men)	approx. 400 days
Voyaging time, Mars to earth	260 days
Total available payload in circum-Martian orbit	approx. 600 t
Total available payload on surface of Mars	149 t
Number of space ships	10
Number of landing boats	3
Required number of ferry vessels	approx. 46
Required number of ferry flights	950
Required time on ferrying operation	approx. 8 months
Propellant cost for ferrying operation	approx. 500 million dollars
Propellant cost for Mars expedition proper	approx. 3.5 million dollars

POWER PLANT PERFORMANCE

*The assumptions heretofore made concerning discharge veloci-
ties of gases and specific performance of the power plants will
now be substantiated in greater detail.*

42. THERMODYNAMIC PERFORMANCE

The quotient $I = F/\dot{W}$ (sec), being the ratio of thrust to
propellant consumption per second, is referred to as "specific
impulse" and is an excellent gauge of the efficiency of a rocket
motor. When the nozzle exit pressure is equivalent to the
ambient pressure, the simple ratio $u = 981 \cdot I$ (cm sec^{-1}) exists
between I and the exhaust velocity u.

"Theoretical specific impulse" is defined as the specific
impulse when there is fully balanced combustion equilibrium
in the chamber and when expansion is isentropic. It is computed as follows:

$$I_{th} = 9.323\sqrt{i_0 - i_e} \tag{42.1}$$

where i_0 is the enthalpy per unit mass of the propellant
combination and i_e is the enthalpy in the nozzle exit after
isentropic expansion.

The stoichiometric equation for the reaction of hydrazine
with nitric acid, without taking into consideration any dissociation, reads:

1.25N$_2$H$_4$ (liquid) + HNO$_3$ (liquid) → 3H$_2$O (gaseous)

$$+ 1.75N_2 \text{ (gaseous)}; \Delta i = -1424.5 \text{ cal gm}^{-1}. \tag{42.2}$$

Here Δi represents the increase in enthalpy which is equivalent
to the negative heat quantity released. It is, nevertheless,
necessary to consider dissociation of the products of combustion at temperatures above 2,000° K, which prevail in
this case. This is done in the following manner:

Compute the dissociation equilibria and their appropriate total enthalpies (chemical and temperature enthalpies) for an arbitrary series of temperatures in the region of the anticipated temperature of combustion T_i, as, for example, $T = 3{,}000°$, $2{,}900°$, and $2{,}800°$ K. By comparing enthalpies thus computed with the total available enthalpy of the propellants as found, using the basic reaction (42.2), it is possible to find the actual temperature of combustion by interpolation.

Thus $T_i = 2{,}850°$ K was found for a combustion chamber pressure $p_i = 15{,}000$ gm cm^{-2} and the equation for the actual process of combustion becomes

$$1.25N_2H_4 + HNO_3 \rightarrow 2.624H_2O + 0.23280OH + 0.2413H_2$$
$$+ 0.0630O_2 + 0.01741O + 0.04156H + 1.75N_2. \quad (42.3)$$

We now determine the entropy S_0 of the gaseous mixture at temperature T_i and combustion chamber pressure p_i. The expansion being isentropic, the same entropy must exist in the exit area of the nozzle, i.e., $S_e = S_0$.

We may now compute the entropies for any given nozzle exit pressure and a number of arbitrarily assumed exit temperatures. The actual temperature in the exit area is obtainable by comparing these entropies against S_0 and interpolating. The enthalpy i_e in the nozzle exit area is then easy to calculate. When exit temperatures are under $2{,}000°$ K, dissociation may be neglected. When i_0 and i_e are thus known, equation (42.1) gives the theoretical specific impulse and thereby the theoretical exhaust velocity. The term "theoretical" is applied to these two figures because the following factors are neglected in their computation:

1. Incomplete combustion on account of defective atomization.

2. Incomplete re-association of dissociated molecules.

3. Friction of gas against nozzle surfaces.

4. Turbulence losses.

5. Divergence losses (there is a velocity component at right angles to the direction of thrust, which, although small, delivers no useful thrust).

The above losses may be combined in a loss factor χ and applied to the calculation. Well-designed rocket motors with nozzles of moderate expansion ratios have loss factors as good as $\chi = .95$, while nozzles with large expansion ratios rarely if ever exceed $\chi = .90$.

Table 32 contains data of the above nature applicable to operating conditions of rocket motors within the scope of this investigation. Table 32 refers to hydrazine and nitric acid in equivalent ratio.

TABLE 32. SPECIFIC IMPULSES AND EXHAUST VELOCITIES

		1*	2	3	4	Dimension
Nozzle exit pressure	p_e	700	24.1	10	66	gm cm^{-2}
Expansion ratio	p_i/p_e	21.4	622	1,500	227	. . . . .
Nozzle exit temperature	T_e	2,005	1,165	983	1,415	°K
Theoretical specific impulse	I_{th}	242	307	318	291	sec
Theoretical exhaust velocity	u_{th}	2,374	3,010	3,120	2,850	m/sec
Loss factor	χ	0.95	0.93	0.90	0.93	. . . . .
Specific impulse	I	230	285	286	271	sec
Exhaust velocity	u	2,250	2,800	2,810	2,650	m/sec

* Column 1 refers to the first stages of the ferry vessels, column 2 to their second stages, column 3 to both interplanetary ships and third stages of ferry vessels, column 4 to landing boats.

43. STATIC THRUST FORCES

The specific impulses shown in Table 32 are valid when it is assumed that the ambient air pressure is equal to the pressure in the nozzle exit area. The latter is determined by the combustion pressure and the fixed ratio between throat and exit area. When the nozzle exit pressure becomes greater or less than the ambient pressure, so-called "static thrust forces" increase or decrease the "nominal" thrust computed from factors above considered. Static thrust is the product of the exit area of the nozzle A_e and the difference between nozzle exit pressure p_e and ambient pressure p_a,

$$\Delta F = A_e \, (p_e - p_a). \tag{43.1}$$

It may be negative or positive, according to whether the ambient pressure is greater or less than the nozzle exit pressure. Table 33 shows the gains or losses via static thrust forces as applying to the more important operating conditions considered in this study.

TABLE 33. GAINS AND LOSSES VIA STATIC THRUST FORCES

	$A_e \cdot 10^{-3}$ (cm²)	p_e (gm cm⁻²)	p_a (gm cm⁻²)	Static thrust force ΔF (absolute)	Percentage gain or loss
Ferry vessel					
First stage:					
Take-off	2,240	700	1,000	$-673 \cdot 10^6$ gm	5.25 loss
Cut-off	2,240	700	4	$1,560 \cdot 10^6$ gm	12.20 gain
Second stage:					
Ignition	3,000	24.1	4	$60.3 \cdot 10^6$ gm	3.77 gain
Cut-off	3,000	24.1	0.2	$71.6 \cdot 10^6$ gm	4.48 gain
Third stage:					
Ignition	740	10	0.2	$7.25 \cdot 10^6$ gm	3.63 gain
Cut-off (and all maneuvers in empty space)	740	10	0	$7.40 \cdot 10^6$ gm	3.70 gain
Space ship	740	10	0	$7.40 \cdot 10^6$ gm	3.70 gain
Landing boat					
Take-off (re-ascent)	176	66	84	$-3.17 \cdot 10^6$ gm	1.58 loss
Cut-off (re-ascent)	176	66	0.2	$11.6 \cdot 10^6$ gm	5.80 gain
Maneuvers in empty space	176	66	0	$11.6 \cdot 10^6$ gm	5.80 gain

44. BASIC DATA OF CALCULATIONS

The exhaust velocity from Table 32, $u_I = 2,250$ m/sec, was used in numerically integrating the ascent track of the first stage of the ferry vessels, the static thrust forces as found from equation (43.1) being appropriately added or subtracted at each step of the computation. Aerodynamic drag was considered separately.

Static thrust forces were neglected in all subsequent cases. Table 33 shows that (with the exception of the first few seconds when the landing boats are leaving Mars) they produce a thrust gain without increasing propellant consumption. Thus neglecting the static thrust forces actually increases the propellant reserves by between 3 and 6 per cent above those specifically referred to during the pertinent performance computations.

The discharge velocities used for computations are as follows: second and third stages of ferry vessels $u_{II} = u_{III} = 2,800$ m/sec; space ships $u = 2,800$ m/sec; landing boats $u = 2,600$ m/sec.

INTERPLANETARY RADIO COMMUNICATION

We shall now discuss the maximum ranges of present-day radio in vacuo, and the antennae and power required to maintain uninterrupted communication between earth and an expedition to Mars. An intermediary satellite vehicle radio station circling in the orbit of departure is postulated to avoid difficulties imposed by earth's rotation and atmospheric effects.

45. EVALUATION OF THE EXPERIMENTAL RADIO CONTACT WITH THE MOON

In 1946 American scientists for the first time succeeded in transmitting radio signals to the moon. These were echoed from the moon and received and recorded on earth. The moon intercepted only a very small quantity of the directed intensity of the transmitter, and only a fraction of the energy actually intercepted was reflected. The power of the "moon transmitter" corresponding to the actually reflected power would have been low indeed. It is therefore obvious that a powerful transmitter in space would have no trouble being received on earth, even at a distance many times that between earth and moon.

Let P be the power beamed at the moon, and let $r \cong 4 \cdot 10^{10}$ cm be the distance between moon and earth. Let $A = 9.5 \cdot 10^{16}$ cm² be the projected area of the moon facing earth. Then the power impinging on the lunar surface facing earth will be

$$P_M = \frac{P \cdot A}{4\pi r^2} \cong 5 \cdot 10^{-6} P. \qquad (45.1)$$

The moon will isotropically reflect but 20 per cent of this power: $P_{refl} = 1 \cdot 10^{-6} P$. It is only one-millionth of the transmitted output power beamed at the moon. Therefore a transmitter possessing the same beamed power as the one used for the experiment would, since power received is inversely proportional to the square of the distance, produce as strong a signal on earth as that reflected from the moon if the transmitter were one thousand times as distant as the moon. Thus it is apparent that the equipment used in the lunar radio experiment would suffice for one-way interspace radio traffic over a distance of $1,000r = 4 \cdot 10^{13}$ cm. This distance is roughly equivalent to the maximum possible distance between Mars and earth ($3.77 \cdot 10^{13}$ cm) existing when Mars stands behind the sun. Thus there can be no doubt that radio communication is range-wise possible to and from Mars at any time.

46. MAXIMUM RADIO DISTANCE WITH PRESENT EQUIPMENT

It is generally accepted that the range of a transmitter increases with increasing output power, higher directivity of transmitting and receiving antennae, and decreasing noise level against which the incoming signal must compete.

a. Available transmitting power

Table 34 shows what has already been done in practice in the matter of transmitter output power at high frequencies.

TABLE 34. TRANSMITTER OUTPUT POWER ACHIEVED IN PRACTICE

Frequency f (megacycles per sec)	Wavelength λ (cm)	Method	Continuous wave power P (watt)	Peak power P (watt)
100	300	Triode, amplifier (lunar experiment)	$15 \cdot 10^3$	. . . .
350–500	85–60	Magnetron	$10 \cdot 10^3$	. . . .
340–625	88–48	Resnatron	$60 \cdot 10^3$	. . . .
500–10,000	60– 3	Magnetron	$1 \cdot 10^3$	. . . .
1,000	30	Magnetron	. . . . . .	$1,000 \cdot 10^3$
3,000	10	Magnetron	$10 \cdot 10^3$	$2,500 \cdot 10^3$
10,000	3	Magnetron	. . . . . .	$1,000 \cdot 10^3$
30,000	1	Magnetron	. . . . . .	$50 \cdot 10^3$

We do not list the interestingly high performances obtainable at lower frequencies. It does not appear expedient to utilize such frequencies for space radio, for reasons subsequently to be developed. Unusual output power is developed by the Resnatron at wavelengths of 88 to 48 cm, and the Magnetron in the 10 cm band (S-band). Both devices were developed by intensive research during World War II.

b. Directional antennae

Directional transmitting and receiving antennae have increasing directivity the greater their physical dimensions are with respect to the wavelengths handled. This more than compensates the higher output power available with long-wave equipment with its practical impossibility to concentrate the energy along a narrow beam (outsize antennae required!).

A directional transmitting antenna having a radiating area of A produces a gain in the receiving antenna, as compared to a spherical radiator, by the factor G.

$$G = \frac{A \cdot 4\pi}{\lambda^2} \cdot \eta \qquad (46.1)$$

where $\eta = 0.75$ is the "antenna utilization factor" and λ the wavelength.

The power intercepted by the receiver is determined by the area of the receiving antenna. If the latter is fixed, it is independent of the wavelength. For any given distance and wavelength, and in view of equation (46.1), it becomes evident that the received signal power is proportional to the product of the areas of the transmitting and receiving antennae. The power input in the receiver remains unchanged if transmitting and receiving antennae be interchanged.

The power gain G of the transmitter with a given antenna area A is inversely proportional, according to equation (46.1), to the square of the wavelength. Thus if the antennae areas are restricted by practical considerations, best results are obtained with short wavelengths.

Let the maximum physical area of a space antenna be $A = 10^6$ cm^2. If such an antenna should be inaccurately constructed to the extent of one thousandth of its linear dimensions, it would affect a 10 cm wavelength to the extent of one tenth of a wavelength. Table 34 indicates that the 10 cm wavelength can give performances not obtainable with shorter ones, thus making it inadvisable to consider them.

The beam width of the 10^6 cm^2 antenna when used with a 10 cm wave and when it is in the shape of a circular parabolic mirror is as follows:

$$\vartheta = \frac{\lambda}{D} = \lambda \sqrt{\frac{\pi}{4 \cdot A}} = 0.855 \cdot 10^{-2} \text{ radians} = 0.5°. \quad (46.2)$$

c. Noise level

Thermal agitation noise of an antenna directed towards empty space is negligible because the temperature of the space at which the beam is directed is only slightly above absolute zero. Such an antenna will, however, pick up the so-called "cosmic noise." The mean intensity of this noise as averaged over the celestial sphere roughly obeys the following formula

$$N_c = 1.8 \cdot 10^6 / f^3 \quad \text{[watts]} \quad (46.3)$$

where f is the frequency in megacycles per second. Equation (46.3) holds for the frequency band between 18 and 200 mc, and at higher frequencies the cosmic noise is negligible.

The noise level of the receiver is determined by the following equation:

$$N_r = \overline{NF} \cdot k \cdot T \cdot B \quad \text{[watts]} \quad (46.4)$$

where $\overline{NF}$ is the noise factor, $k \cdot T = 4.11 \cdot 10^{-21}$ joules (at $T = 300°$ K), and B is the bandwidth of the receiver expressed in cycles per second. Table 35 shows values for the noise factor. It will be noted that, according to equation (46.4), the noise level of the receiver increases with the bandwidth.

TABLE 35. NOISE FACTOR

Frequency f (mc)	Method	Noise factor, $\overline{NF}$
50	"Lighthouse"	2.1
100	Tube amplifier	2.5
200	Tube amplifier	3.0
600	Tube amplifier	5.0
1,000	Crystal	10
10,000	Crystal	10

With respect to the noise alone, the best frequency for space radio communication lies at the point where the sum of receiver noise (46.4) and cosmic noise (46.3) is a minimum. This minimum occurs between frequencies of 100 and 200 megacycles per second, i.e., between 300 and 150 cm wavelength. Frequencies below 100 megacycles are unsuited to space radio because of the strongly increasing cosmic noise. Above 200 megacycles, receiver noise predominates while cosmic noise becomes negligible.

d. Increase of range by pulse transmission (radar)?

Transmitter output power may be very large during short pulses if care is taken that intervals between pulses are long enough to permit the power loss (plate dissipation) to remain heat-wise within acceptable limits for the tube. For example, a tube built for 10^3 watts continuous wave operation will withstand as much as 10^6 watts, if such peaks are applied 1,000 to the second and each pulse lasts but one micro-second (see Table 34).

However, optimum reception of pulses of a duration τ calls for a receiver bandwidth of

$$B = \frac{1}{\tau}.$$ (46.5)

In the case of greatly boosted power output by the use of short pulses the average energy given off by the transmitter is approximately constant:

$$E = P \cdot \tau \cong \text{const.}$$ (46.6)

In view of equations (46.5) and (46.6), the useful signal power therefore is

$$P_r = \text{const} \cdot B$$ (46.7)

which is to say that the signal power received is proportional to the bandwidth.

But according to equation (46.4), the noise power in the receiver is also proportional to the bandwidth. According to the assumption made in equations (46.5) and (46.6), the signal-to-noise ratio (in first approximation) becomes thus independent of a power boost via pulses. The conclusion is that the allurement of high peak power available in pulse-radar is a snare and a delusion and not conducive to the achievement of extreme ranges.

e. Narrow-band operation

The noise power in the receiver recedes in the same ratio as the bandwidth is reduced, according to equation (46.4). Thus, in the theoretically limiting extreme of a zero bandwidth, the noise power would be zero also. The transmitter would then be perceptible (though without signal communication) at any arbitrary distance, because its signal would have no noise to overcome.

In practice, bandwidths cannot be narrowed beyond certain limits. These limits will be treated below. The rapidity and versatility of radio communication attainable with these smallest possible bandwidths emerge pretty much automatically.

f. Frequency constancy

The bandwidth of the receiver must be wide enough to receive the carrier frequency of the transmitter, despite its fluctuations and those of the receiver's intermediate frequency. Quartz crystals of a constancy of 10^{-8} are already in practical use and there is no reason why they should not be used in space ships. Let it therefore be assumed that crystals of this frequency stability will be available to stabilize the transmitter frequency and also will be utilized in the receiver.

Let us further assume that a high-powered transmitter of inherent stability of 10^{-4} has been improved to 10^{-7} by automatic frequency control with such a 10^{-8} crystal oscillator. In the receiver, intermediate frequencies of a stability of 10^{-8} may be available. The minimum required bandwidth of the receiver over a series of working frequencies may be taken from the third column of Table 36.

TABLE 36. MINIMUM BANDWIDTH REQUIRED

Frequency f (mc)	Maximum difference frequency caused by fluctuations of transmitter and receiver frequencies (cycles per sec)	Minimum bandwidth B_{min} (cps)
100	10	20
300	30	60
600	60	120
3,000	300	600

If no special precautions are provided for suppressing the image frequency, it will, of course, be necessary to double the minimum bandwidths derived in Table 36. Thus the band-width determining the noise power becomes

$$B_N = 4 \cdot 10^{-7} \cdot f. \qquad (46.8)$$

It is therefore plain that the noise power, proportional to B_N, increases linearly with the working frequency by reason of imperfections in frequency stability.

The Doppler effect will produce marked frequency shifts between transmitter and receiver in interplanetary radio, which will be considered in greater detail in section 48. These shifts have no bearing on the minimum bandwidth required. Frequency changes by Doppler effect take place so slowly that current tuning of the receiver by automatic frequency control should be relatively easy, unless extremely narrow bandwidths are employed.

g. Selection of frequency for maximum range

The useful receiver input is

$$P_r = \frac{P \cdot G \cdot A \cdot \eta}{4\pi r^2} \text{ [watts]}. \qquad (46.9)$$

The range r reaches its maximum value r_{max} when this signal power input equals the noise power of the receiver, referred to its input. Using equation (46.4) we then have:

$$P_{r,min} = \frac{P \cdot G \cdot A \cdot \eta}{4\pi r^2_{max}} = \overline{NF} \cdot k \cdot T \cdot B \qquad (46.10)$$

where $A \cdot \eta = A_c$ is the collecting antenna area which is in-dependent, in first approximation, of the frequency. G is found by the use of equation (46.1), and we may generally put $G = \text{const}/\lambda^2 = \text{const } f^2$. Equation (46.8) shows generally that $B = \text{const } f$. Equation (46.10) then yields the general relationship

$$r_{max} = \text{const} \sqrt{\frac{P}{\overline{NF}} \cdot f} \;. \qquad (46.11)$$

If we now introduce into (46.11) the available transmitter output powers P as shown in Table 34 (continuous wave power) and the noise factors $\overline{NF}$ as listed in Table 35 for the different frequencies f, it is found that there is no particularly outstanding optimum frequency for maximum range r_{max}. Within the frequency range of the Resnatron, range increases slightly, so that the maximum distance is found for a frequency of 600 megacycles. However, the range obtained with 3,000 megacycles is but slightly shorter despite the transmitter output being but one sixth of that at 600 megacycles.

Thus the absolutely maximum range can be obtained with a transmitter of the Resnatron type in the 50-cm band (see below), but in actuality the 10-cm band is superior to it by reason of, first, the lower power required and, second, the inherently wider bandwidth (Table 36).

h. Maximum range

We shall below examine the maximum range of a transmitting and receiving reciprocality which will generally approach, with respect to antenna size and transmitter output, the maximum obtainable with present radio gear. We make the following assumptions for the equipment:

Frequency: $f = 600$ megacycles per second (Resnatron).

Wavelength: $\lambda = 50$ cm.

Transmitter output power: $P = 6 \cdot 10^4$ watts (continuous wave).

Antennae areas: $A = 10^6$ cm^2 (physical size, equal for receiver and transmitter).

Collecting antennae areas: $A_c = A \cdot \eta = 7.5 \cdot 10^5$ cm^2.

Receiver noise factor: $\overline{NF} = 5$.

Receiver bandwidth: $B = 120$ cycles per second (image frequency suppressed).

Equation (46.10) then yields

$$ r_{max} = \sqrt{\frac{P \cdot G \cdot A_c}{4\pi \cdot \overline{NF} \cdot k \cdot T \cdot B}} . \qquad 46.12) $$

From the above assumptions, the limiting range for the equipment is

$$ r_{max} = 2.35 \cdot 10^{15} \text{ cm.} \qquad (46.13) $$

Pluto, the outermost planet of the solar system, has an orbital diameter of $1.2 \cdot 10^{15}$ cm as a matter of comparison. This range is the absolute extreme which could be covered by present equipment. This is to say that signals received at this range could *perhaps* be sorted out—if given in the form of long, slow dots and dashes—from the equally strong noise or "grass," and be barely detectable on the screen of an oscilloscope.

47. DESIGN OF SUITABLE RADIO EQUIPMENT
FOR THE MARS EXPEDITION

The limiting range for radio equipment as selected in the preceding section exceeds the maximum possible distance between earth and Mars ($3.77 \cdot 10^{13}$ cm) by a factor of 62. Performance-wise this means that an excess of 36 decibels will be available, even when Mars is most distant. When Mars is in opposition ($77 \cdot 10^{11}$ cm) the excess would amount to 49 decibels.

The reserve range thus available with respect to radio communication with an expedition to Mars may be exploited in the following manners:

1. Reduction of transmitter output power, thus reducing transmitter weight and size of the solar reflector for power generation.

2. Reduction in size of the directional antennae.

3. Increasing the bandwidth, thus rendering possible reception and transmission of speech and music.

Which of these possibilities will be principally exploited will depend upon how much effort one desires to expend to increase the range of radio telephone communication. The radio station of a space ship treated below was selected so that its total weight would not be over 3 metric tons, including solar power plant and storage batteries. Thus it may be carried by one of our "cargo vessels." The antenna area was limited to 10 m². A similar station would encircle earth in the orbit of departure, except that its antenna area would be 40 m².

The main characteristics of the suggested equipment are as follows:

Frequency: $f = 3,000$ megacycles per second (Magnetron).

Wavelength: $\lambda = 10$ cm.

Transmitter output power: $P = 10^4$ watts (ship station and orbital station).

Antenna area (ship station): $A_s = 10^5$ cm².

Collecting antenna area (ship station): $A_{c,s} = 7.5 \cdot 10^4$ cm².

Antenna area (orbital station): $A_o = 4 \cdot 10^5$ cm².

Directivity of this antenna: $G_o = 3.78 \cdot 10^4$ (See Eq. 46.1).

Receiver noise factor: $\overline{NF} = 10$ (See Table 35).

If we further assume that the useful signal power in the receiver input must exceed the noise power by a factor of $a = 100$ (20 decibels excess) for satisfactory speech transmission, and that for the latter a minimum bandwidth of $B = 5,000$ cycles per second is required, then equation (46.12) shows that the above combination will have a limiting range for speech transmission of

$$r_{max} = \sqrt{\frac{P \cdot G_o \cdot A_{c,s}}{4\pi \cdot a \cdot \overline{NF} \cdot k \cdot T \cdot B}} = 1.05 \cdot 10^{13} \text{ cm} \qquad (47.1)$$

The expedition would reach this limiting range after about 160 days of Marsward travel. Equation (47.1) is based upon the assumption that means have been provided for suppressing image frequencies (see section 46f). Such means are relatively simple and effective.

When it is no longer possible to maintain radio telephone communication by reason of increasing distance, automatic telegraphy can be used. If the bandwidth required therefor is $B = 1,000$ cps, and if the signal-to-noise ratio is reduced to $a = 20$ (13 decibels), the limiting range becomes

$$r_{max} = 5.25 \cdot 10^{13} \text{ cm}. \qquad (47.2)$$

Hence automatic telegraphy is always possible at any distance between earth and Mars, since the maximum distance is $3.77 \cdot 10^{13}$ cm. The range is, of course, even greater when hand key is used. This is due to the modest bandwidth and signal-to-noise requirement of slow dot-and-dash telegraphy with acoustic or optical signal registration.

48. DOPPLER EFFECT

Any relative movement between transmitter and receiver taking place in the direction of wave propagation produces a Doppler effect, i.e., a difference between transmitted and received frequencies. Doppler effects of considerable magnitudes are produced between the Mars vessels under way and the radio station circling in the orbit of departure by the following:

1. Orbital motion of the earth station.
2. Relative motion between ships and earth on both Marsward and earthward ellipses.
3. Orbital motion of ships around Mars during "waiting time."

4. Relative motion between earth and Mars during ships' stay in the circum-Martian orbit.

The magnitude of any Doppler effect is computed as follows:

$$\Delta f = \frac{\Delta v}{\lambda} \qquad (48.1)$$

where Δf is the frequency shift, Δv the relative velocity between transmitter and receiver in the direction of wave propagation, and λ the wave length. The orbital movement at circular velocity $v_{ci,1} \cong 7 \cdot 10^5$ cm sec^{-1} of the earth station produces, for example, a maximum variation in the relative velocity during an hour's communication of $\Delta v = 14 \cdot 10^5$ cm sec^{-1}. Considering a wavelength of $\lambda = 10$ cm, a Doppler effect of

$$\Delta f = 140 \text{ kilocycles per second} \qquad (48.2)$$

ensues. The maximum Doppler effect occurs when sun, earth, and Mars form a right angle. In that case, very nearly the whole of earth's circum-solar velocity becomes a factor and the Doppler shift becomes

$$\Delta f = \frac{v_E}{\lambda} = 298 \text{ kilocycles per second.} \qquad (48.3)$$